Gertjan Dijkink

TERRITORIAL SHOCK
The Moral Impact of Boundary Change
in Two Millennia

LIT

Cover image: Historically ambiguous boundaries. Relief map of the Pyrenees mountains (Cerdagne). See figures on page 77.

This book is printed on acid-free paper.

Bibliographic information published by the Deutsche Nationalbibliothek
The Deutsche Nationalbibliothek lists this publication in the Deutsche Nationalbibliografie; detailed bibliographic data are available on the Internet at http://dnb.d-nb.de.

ISBN 978-3-643-91012-7 (pb)
ISBN 978-3-643-96012-2 (PDF)

A catalogue record for this book is available from the British Library.

© LIT VERLAG GmbH & Co. KG Wien,
Zweigniederlassung Zürich 2019
Klosbachstr. 107
CH-8032 Zürich
Tel. +41 (0) 44-251 75 05
E-Mail: zuerich@lit-verlag.ch   http://www.lit-verlag.ch
Distribution:
In the UK: Global Book Marketing, e-mail: mo@centralbooks.com
In North America: Independent Publishers Group, e-mail: orders@ipgbook.com
In Germany: LIT Verlag Fresnostr. 2, D-48159 Münster
Tel. +49 (0) 2 51-620 32 22, Fax +49 (0) 2 51-922 60 99, e-mail: vertrieb@lit-verlag.de

e-books are available at www.litwebshop.de

Gertjan Dijkink

Territorial Shock

Contents

Preface i

1. Territorial Shock: an introduction 1

2. Barbarians at the gates: the classic empires 25

3. A new Jerusalem: the birth of the territorial state 53

4. The vertigo of public space (High Modernity 1815-1980) 89

5. Can the centre hold? Territory in the age of Late Modernity ... 121

6. Dying states: prelude to re-territorialization? 157

7. Globalization and its detractors 207

    Appendix: the argument in brief 225

    Literature 229

    Index 245

Preface

A substantial part of this book was written in the first decade of the new millennium, an era that dawned with a bang rather than a whisper. Two major shocks did not bode well for the near future (as later international entanglements have confirmed): the terrorist attacks on New York's WTC towers in 2001 and the financial crisis starting in 2007. These were territorial shocks because they induced a disruption in our mental maps of the world: our assumptions about the distribution of power and security of borders and areas. The mentioned events first cast doubt on the power of sovereign states to guarantee the safety and wealth of their citizens. Second, they suggested a weakness of Western social systems, the US in particular, and a shift of power to other parts of the world (like Asia). Students of international relations were already discussing such issues during the last quarter of the 20th century. I have vivid memories of a 1999 conference entitled 'Challenging the American Century' (directed by political geographer Peter Taylor in Loughborough) in which doubts were raised about the US remaining the dominant power in the 21st century. Although many participants did their best, perhaps desperately, to identify the enduring assets of the American system. Anyhow, dramatic real-world events unleash a much wider chain of reactions by actors who are not particularly interested in well-considered academic advice. Restoring the familiar order is always the first incentive, and if that fails, one needs to infuse the changed world with a new moral satisfaction.

The changing territorial order and the moral challenge of a globalizing world (the new vulnerability of states) are the subject of this book. The territorial challenges of our epoch are not unique. Earlier in history people have been plunged into territorial 'revolutions', although never instantaneous, that required a new mental and moral map of the world. The current book originated in the study of such shifts, a subject relatively neglected in political science and history in the 20th century. The Department of Geography Planning and International Development Studies (GPIO) at the University of Amsterdam offered freedom and forum to develop and discuss such theoretical explorations. I am also grateful for a grant from NWO (the Netherlands Organization for Scientific Research, research programme 'Shifts in Governance') in 2004 for this part of the research, entitled at the time 'Planting the Flag'. Since that date others have taken up the thread and produced brilliant studies, like Stuart Elden in geography and Charles Maier in political history[1]. My own interest, however, shifted to the relationship between impersonal geopolitical realities and human experience, particularly the moral justification or legitimacy of authority and boundaries. That perspective was also useful in linking the abstract theme of territoriality to the life-world of students in courses on political geography at the University of Amsterdam. This quest is also a sequel to my earlier work on geopolitics, identity and religion.[2] An early article on the theme of this book appeared in the French web journal *L'Espace Politique* (2010).[3]

The study that appears here deals with the history of territorial authority as a series of disturbing geopolitical 'earthquakes', the territorial consequences of globalization, the collapse of (dying) states as an opportunity to adapt to a new era and the moral challenge of the current era. The reader will notice that music appears several times in this work as a bridge between local distress and the established (territorial) order. The 16th century huntsman who, unaware of the presence of the Spanish army, chased them away from the Dutch woods by blowing the national hymn on his horn; the Meistersinger of Nuremberg who needed liberation from the narrow conventions of their time; the galvanizing effect of pop music in Estonia's national independence struggle; Bruce Springsteen's lamentation about post-industrial trauma: these examples all attest to music being part and alleviation of territorial shock.

Music is liberating because it transcends boundaries of space and time. I can only add my own experience of glocalization as a member of a chamber music ensemble with changing French and German (and myself Dutch) participants in Southern France, the area bordering (French and Spanish) Catalonia. The ensemble smoothly linked the international with the local, enabling a much deeper exchange of cultural *and* political views among both musicians and inhabitant than our dominant territorial tradition would allow.

---

[1] Stuart Elden, 2013, *The Birth of Territory*. Charles S. Maier, 2016, *Once Within Boundaries. Territories of Power, Wealth and Belonging since 1500*. See for earlier introductions David Delaney, 2005, *Territory: A Short Introduction*. and David Storey, 2001, *Territory: The Claiming of Space*.
[2] Gertjan Dijkink, 1996, *National Identity and Geopolitical Visions. Maps of Pride & Pain*. Gertjan Dijkink, 2006, 'When religion and geopolitics fuse'.
[3] Gertjan Dijkink, 2010, 'Territorial Shock: toward a theory of change'. *L'Espace Politique* 12(3).

This work has also benefitted by a dialogue of many years with interested students in political geography at the University of Amsterdam. One of them, Tessa Wernink, even started to correct the workbook of my lectures in order to facilitate a future publication. The finalizing of this book has been relieved by the help of three dedicated readers. First my colleague Virginie Mamadouh has pointed out inconsistencies and gaps in my approach to the subject of territorial shock. Of course, it does not mean that she agrees with any opinion or approach in this book. I have many dear memories of our long-standing collaboration. Mandy Hoggard produced a high-quality language revision and never tired of indicating ambiguous sentences or paragraphs. I felt privileged to receive her help. Finally, my wife, Eva van Kempen, always observed when interesting stories diverted the text from the central argument and consequently urged me to 'kill my darlings'.

Gertjan Dijkink
Amsterdam, September 2018

Chapter 1
TERRITORIAL SHOCK: AN INTRODUCTION

> Earthquake survivors often report a severe sense of disorientation during severe groundshaking, a whole-body confusion and loss of moorings - both figurative and literal. Our frame of reference is the ground we stand on, and when that reference frame shifts without warning, something deep in us is affected. The earth simply isn't supposed to move, we assume, and when it does we are profoundly unsettled.
> Earthquake consultant **Richard Briggs** in *Asia Times*, October 9, 2009

> He established the earth on its foundations; it will never be shaken.
> **Psalm** 104:5 (Christian standard Bible)

Geopolitical earthquakes and dissolving boundaries

'From one moment to another…it is terrible because you don't have the physical place of memories... I knew only that I could not come back home … Therefore, life stopped'. Such were the accounts of helpless victims when asked to describe their feelings in the aftermath of an Italian earthquake.[1] It is striking that when faced with an event of such magnitude, people do not primarily recall the horror of possible material loss and human death, but rather the disruption of space and time. The loss of one's frame of reference and the inability to imagine the continuation of life is a well-known experience of earthquake victims. In Elena Ferrante's four-novel epic about growing up in post-war Naples, the narrator (Elena, one of the two central characters) describes the experience of the Eboli earthquake: 'The earthquake – the earthquake of November 23, 1980, with its infinite destruction – entered our bones. It expelled the habit of stability and solidity, the confidence that every second would be identical to the next… A sort of suspicion of every form of reassurance took over, a tendency to believe in every prediction of bad luck, an obsessive attention to signs of the brittleness of the world, and it was hard to take control again'.[2] For Lina, the other main character, the experience of brittleness assumes a pathological shape when she witnesses

---

[1] Barbara Lucini, 2014, *Disaster Resilience from a Sociological Perspective: Exploring Three Italian Earthquakes as Models for Disaster Resilience Planning*, p. 162.
[2] Elena Ferrante (Ann Goldstein trans.), 2015, *The Story of the Lost Child*. Ch. 50.

people fleeing in cars: 'Gasping for breath, she cried out that the car's boundaries were dissolving, the boundaries of Marcello, too, at the wheel were dissolving, the thing and the person were gushing out of themselves, mixing liquid metal and flesh. She used that term: "dissolving boundaries"'.[3]

There is no mistaking that the perception of 'dissolving boundaries' is linked with a condition of extreme fear induced by external events but possibly also by personal characteristics, as in Lena's case. Real-world events erase physical and social boundaries, but they may also cause an internal process in our minds as well. From the refugee crisis in Europe to the out-migration of employment from the United States (the end of the American Dream), people are anxious to save or recover what they understand as their proper home, if necessary by erecting walls. The common factor is 'territorial shock', the experience that people, activities and institutions have lost their ordained place, resulting in a high level of unpredictability. While there are scarce solutions for nature-made territorial shocks like earthquakes, man-made territorial shocks often unleash dramatic political reactions, for example erecting walls. If that is to no avail, people may seek salvation in religion or by shifting their identity and values.

As the Psalmist proclaims, 'He established the earth on its foundations; it will never be shaken'. But when the earth nevertheless trembles we have two options: abandon our belief or search for a religious interpretation that again restores order to the world. Indeed, it is not surprising that people seek comfort in the eternal truths of religion. For example, a sudden appeal of religion was recorded after the earthquake that destroyed Christchurch, New Zealand in February 2011.[4] The geophysical forces that cause earthquakes have their counterpart in the human world in geopolitics or geo-economics. Geopolitics is about the way the earth is used (as bounded space or physical resource) in developing or defending one's power. The state is still

---

[3] Ibid. Ch. 51. It should be mentioned that the translator selects a more geographic term (dissolving boundaries) than appears in the Italian original where Ferrante uses the not very common verb *smarginare* which means so much as 'eliminating the margins (of a printed text or photograph') However, the meaning of 'merging' distinct surfaces or figures (or more precisely erasing outlines) remains intact in the translation.

[4] Chris G. Sibley and Joseph Bulbulia, 2012 'Faith after an earthquake: a longitudinal study of religion and perceived health before and after the 2011 Christchurch New Zealand earthquake'. Data from before and after the earthquake showed a substantial rise in the number of people who reported to be believers. Ironically, the new converts did not seem to derive much comfort from their faith in contrast to established believers.

the main geopolitical actor and its fortunes are the source of great emotions in times of change. The image of tectonic plates that press toward each other and crush the earth's surface is irresistibly evoked in geopolitics. A simple Google search easily produced more than 3300 results for the phrase 'geopolitical earthquake'.[5] Among them we encounter varied issues, such as: the demise of the communist bloc in the period 1989-1991; the 2007 financial crisis' influence on US hegemony involving and the subsequent rise of a new power bloc, the BRICS (Brazil, Russia, India, China, South-Africa), with their own development bank; and the collapse of Arab states under the influence of Islamic State actions. Such events may indeed disrupt deeply ingrained expectations about how the world works, such as the assumptions that states are more powerful than gangs or that globalization is a process that runs from the US (or the West) to the rest of the world.

It is easy to see why the earthquake has become such a tempting metaphor for events that impinge upon us not from the natural but from the *social* world.[6] We can cope with a lot of change in both public and private human affairs– even with the death of our beloved – but certain changes affect our expectations of how the world works to such an extent that they cause a feeling of *disorientation*. The demise of the communist universe deprived many people in the former Soviet Union of a sense of direction in life.[7] The introduction of capitalism with its moral void caused not only a cultural shock but a territorial shock as well; it seemed to hand people over to unchecked influences from the outside, call it globalization. The state was no longer a sanctuary.

When talking about geopolitical change in these terms, one may easily overlook the subtle distinction between two kinds of long-term territorial change. One, of course, is the disruption of boundaries of influence or power, such as the conquest of another country or the fading away of the Iron Curtain. The other is the supposed demise of the established international system as a system of states that behave as if they are 'closed containers'. The story of territoriality is not merely one of redrawing political boundaries and

---

[5] Accessed October 18, 2014.
[6] Pope Francis' urge for reforms in the Catholic Church in 2014 were called an earthquake in several media. One of the finest: 'Pope Francis' synodal hurricane of love provokes earthquake of mercy'. http://www.nationalreview.com/corner/390198/pope-francis-synodal-hurricane-love-provokes-earthquake-mercy-kathryn-jean-lopez
[7] Svetlana Alexievich, 2016, *Secondhand Time: The Last of the Soviets.*

sending people into exile (geopolitics of the body). That happens just as well today. We may at the same time feel that political boundaries are no longer what they used to be. Institutions and activities such as multinational firms, security agencies and financial streams cross state boundaries unchecked and may even create territorial bases external from internationally recognized states (like Guantanamo prison camp or Jewish settlements in occupied Palestinian territory). This is what we mean with the statement that not only territories have changed, but territoriality as well. The advance of 'globalization' comes with a double attack on our moorings: new boundaries are still created, as in the past, but we feel less safe within them. The territorial shock that is the focus of this book concerns this second meaning of territorial change; We will not focus on boundary actions, but rather on the changing character of the entire territorial practice. This involves three aspects: *closure, governance,* and *identity*.

Territoriality evolved before our 'age of globalization' as well, for example with the emergence of empires or the formation of national states in Europe. Such changes occur more slowly than the liquidation of boundaries in wartime and will not necessarily be recognisable when one's perspective is limited to the human life span. But why shouldn't they reveal themselves in traumatic events like the sudden quaking of tectonic plates from the geophysical realm? Events such as the terrorist attacks of U caused a territorial shock not because the territory was endangered but because it revealed a new type of vulnerability that manifested itself *inside* society rather than at the boundary. During the first hour of the September 11 attacks in the United States, TV reporters used words such as 'Pearl Harbour', 'war', 'God', 'pray' and 'America', suggesting a struggle of historic and cosmic proportions.[8] Appealing to religion and historic-peak experiences is obviously a way to cope with perplexing events that violate our physical security. On the other hand, one may wonder if any sense of discomfort with social change that involves some aspect of boundaries should be defined as 'territorial shock'. This is an ontological question: Is what we call territorial shock not rather economic, cultural, or simply technological ('future'[9]) shock?

There is no doubt that the current political discontent that threatens mainstream political parties in the West springs from a structural economic

---

[8] Amy Reynolds and Brooke Barnett, 2003, '"America under attack": CNN's verbal and visual framing of September 11'.
[9] Alvin Toffler, 1970, *Future Shock* deals with the shock of rapid technological change.

change. This change is characterized by craftsmanship becoming less important than having financial power as investor, middle-class incomes stagnating for decades and labour-intensive parts of production being outsourced abroad, leaving behind a much smaller group of specialists in (hi-tech) communication and services. Likewise, we may attribute the growing discomfort with migration to culture shock rather than territorial shock. The disturbance of many people's mental equilibrium, expressed in protests of 'globalization', would then be classified as an emotional reaction to several crises in the economic and cultural sphere rather than the impact of territorial change. Yet, we may hold on to the term 'territorial shock' in instances where these social changes become dramatized by tracing them back to the problem of 'leaking boundaries', or in President Trump's locution, 'the rape of the nation' by external actors.

A thornier question concerns the nature of the changes that happen around us. Are we indeed living to see an epochal change in the geography of power? Was September 11 the symptom of a subterranean shift connected with a new type of *territoriality*? Or Putin's destabilizing actions in the 'Eurasian' continent? Or the rampant surge of the Islamic State in the Middle East? Such actions seem to resemble alarming concepts like 'geopolitics of chaos' or 'the anarchical society', which were introduced decades ago by authors who foresaw a sweeping change in the organization of power in our age.[10] However disturbing such events are, neither the violation of a boundary nor the relocation of a boundary, nor the rise of a new type of ('hybrid') warfare, necessarily indicate changing territoriality. It is easy to find examples of such seemingly unconventional practices in the past.

Territoriality, or 'territorial order', is a system of rules and practices which sustains an order of territorial 'assemblages' that in its turn has emerged from earlier such orders by a process called re-territorialization.[11] The international state system that started to consolidate in 17th century Europe is an example. It emerged from a territorial order that we may call *imperial*: a junction of different ethnic groups or tribes under the strong military rule of an Emperor who rules over an area with unstable frontiers (Chapter 2). If we compare that with our familiar territorial order, the world of nation-states, some differences immediately come to the fore. The imperial boundary is a

---

[10] Hedley Bull, 1977, *The Anarchical Society: A Study of Order in World Politics*. Ignacio Ramonez, 1996, *Geopolitics of Chaos*.
[11] Deleuze, G. and Guattari, F., 1980, *Mille Plateaux : Capitalisme et Schizophrénie*.

war zone populated by barbarians that are either friendly and welcomed in the empire as warriors, or hostile barbarian tribes that require constant military sorties. Peripheral regions of the Empire, also known as 'marches', could have been former independent kingdoms that retained autonomy because of their value as defensive forces that could swiftly react to aggression from the outside. This whole question of who belongs inside or outside a political unit and its many shades of in-between can be denoted as a matter of *closure*. Another difference between nation-state and empire is that the maintenance of order in the empire is dominated by military operations or coercion rather than by the economic or infrastructural regulation with which we are familiar in nation-states. Such differences will be denoted as the dimension of *governance* in a territorial type. Third, the personal justification for being a subject of imperial rule is the divine quality of the Emperor, whereas in nation-states it is citizen rights. This is the *identity* dimension in our schema of types of territoriality.[12] If we call territory any geographical area with a dominant actor whose actions are based on coercion and/or consent, then territoriality embraces the nature of rules and actions over the territory, namely the treatment of its boundaries (*closure*), the institutional resources used to exercise authority (*governance*) and the way a territory and its authority are incorporated into individual practices and moral judgments (*identity*).

Table 1.1. Variability of territoriality along three dimensions

Dimension	Unbounded (fragmented)	Bounded ('territorial')
Closure Structure of authority. Who are (un)affected by dominant authority?	domains of action gangs diaspora	territorial sovereignty spatial equality
Governance How is a territory / group engaged in turning it into a resource?	personal power / murder blackmail financial services	taxation infrastructural power rule of law
Identity How do people attribute *meaning* to power in (territorial) assemblages?	universalistic ideal religion	nationalism

Since the world of 'closed containers', nation-states, is such a common-sense model of territoriality, it may help to understand other varieties of territorial-

---

[12] I am indebted to the Finnish geographer Anssi Paasi for an earlier version of this triad as part of a general theory of regionalisation. The maturation of any 'region' (or territory) would involve a territorial, an institutional and a symbolic dimension (somewhat equivalent to closure, governance, and identity). Anssi Paasi, 1986, *The Institutionalization of Regions*.

ity by means of the dichotomy of table 1.1. This table shows that all forms of human binding (*closure*), control (*governance*) and satisfaction/meaning (*identity*) may run via (bounded) state-containers, but also in the shape of '*unbounded*' types of organization like gangs, capital (bit-coins) or religion. Consider any type or manifestation of territoriality as a distinct mix of 'choices' from both columns, or as 'deviations' from the common-sense model of political territoriality that essentially corresponds with the last column.

The transition from imperial territoriality to the modern state was a gradual process that never deliberately aimed for the final model that we know today. Each step aroused resistance among those who lost privileges and created confusion among people who just tried to understand their changed place in the world or the cosmos. The concept territorial shock is the epitome of this inability to digest change. But how to trace it and pin it down on specific changes if historic change is such a gradual process? In the following chapters I deal with two pivotal 'moments' in the development of modern territoriality: the transition from imperial to early-modern territoriality (Chapter 3) and the nationalist revolution (high modernity, Chapter 4). The key to understanding the corresponding territorial shocks lies in the shifting role of the three dimensions of table 1.1. They may **actively** contribute to the formation of a political map like imperial *closure* and *governance* but also remain **passive** like *identity* in the same territorial order. Imperial identity implied a static picture of divine authority culminating in the Emperor. It did not legitimate specific imperial boundaries (conquest) or the nature of internal governance. That changed in Europe when kings claimed sovereignty over their 'realms' and the idea of multiple authorities had to be reconciled with the idea of divine support for their rule. Identity became an active component of the territorializing drive in the Early-Modern state. This shift from something implicit into a driving force that amalgamated with *closure* was also the root of early-modern territorial shock, ultimately alleviated by religious change.

Table 1.2 displays an elaboration of the principles explained above for periods that are generally distinguished in European history. Not all authors distinguish a sharp break between the Early- and High-Modern periods but I find the distinction useful in the context of territorial shock. The last period, 'Late-Modernity', corresponds with what is often called the 'Age of Globalization' which is still raging around us. Therefore, its interpretation is more speculative (hence the question mark). The scheme suggests that territorial shocks stemmed from the activation of the *identity* dimension in the Early Modern period and the *governance* dimension in the High-Modern period. The activation of *closure* in our own age should not be taken too literally. Admittedly, more walls are built to stop people than in the previous century, but closure is particularly a challenge concerning the question of *who* should control transnational capital, information and trade (Chapter 5).

Table 1.2 Territorial orders in European history (A=Active, P=Passive)

Territorial order	Closure	Govern.	Identity
Imperial	A	A	P
Early-Modern (1550-	A	P	A
High-Modern (1800-	P	A	A
Late-Modern (1980-	A	A	P?

The scheme of table 1.2 is a heuristic device and does not imply that the author views world history as a mechanical process. However, there is a systemic consistency in each period which fosters dominant preoccupations with social order, while at the same time taking for granted conditions like the existence of an authority that connects heaven and earth or the separation of classes. The change of such silent habits and assumptions demands more of our cognitive and emotional apparatus than matters that are already in a turbulent condition. One should also be aware of the pitfalls of the term 'territorial model' since there was rarely a historic period in which even a part of the world could be described as neatly fitting a territorial model in the sense of a clear design for the exercise of power. Several territorial practices could exist together on the European continent. Even the word model promises a picture with too much detail. To borrow an image from physics: one should follow an approach that is like quantum mechanics (diffuse properties and states of turbulence/activity) rather than Newtonian mechanics (fixed objects) to describe types of territoriality. My aim in distinguishing 'closure', 'governance' and 'identity' is to characterize differences in territorial practice indicated by the fact that each historical period shows an emphasis on dimensions in a state of turbulence rather than on the genesis of a distinct model.

Allow me to enlighten the reader on this point by briefly anticipating the (mainly European) stories told in the next three chapters.

- It starts with the imperial epoch from the ancient Chinese and Roman Empires to Medieval Europe, a period in which central powers were distant from ordinary people and governing the Empire required continuous war against barbarians at the gates and disciplining governors (vassal states) at the margins of the Empire. Obviously 'governance' and 'closure' were in a state of turbulence, but not 'identity'. Most of the unity in medieval Europe was warranted by a stable identity: belief in the divinity of the emperor and the Christian message.

- From the 13th century on, local lords or kings started to strengthen their grasp on the daily activities in their area. Personal gain was the inspiration for this policy, but it also met the common people's need for more safety and regulation of practical conflicts. Kings presented themselves as emperor in their own realm, but this collided with the idea of the divine origin of government still embodied in the twofoldness of Pope and Emperor. This dilemma clearly pulled 'identity' within the sphere of turbulence (Reformation, religious wars). The shock that accompanied the transformation of the European arena into a system of states was predicated on the turbulence of 'identity'. Authority in the sense of sovereignty embodied by a hereditary monarchy became the accepted form of 'governance', shifting it to a more passive role. All turbulence resided in closure and identity. The legitimacy of 'governance' (by a royal family) was not an issue, but boundaries were (and the identity of the people living within them). Where kings lost territory, like in the Low Countries at the course of the 16th century, a feverish search for identification started.

- When the map of Europe as a system of states somewhat stabilized at the start of the 19th century ('closure' passive), the state appropriated a new meaning almost as a kind of factory. Sharply decreasing transport costs (enabled by the invention of the steam engine) boosted the export of industrial products. Trade became an important source of wealth and national power. Now the 'governance' dimension became active in supporting the competitive edge of the state in the international arena. Its specific form, infrastructural power, involved greater numbers of people who in a way became shareholders in the state enterprise and who expected some recognition. The turbulence emerged in a 'governance-identity' compound that exposed people to new public forces and interdependencies (called 'vertigo' in the title of chapter 4). The shock of having to accept foreigners (from different villages, families, or estates) as members of the same collective enterprise produced the well-known movement of nationalism.

Globalization: clash of civilizations or territorial shock?

At the approach of the 21st century (in 1993), political scientist Samuel Huntington raised a lively discussion in and outside academic circles with the thesis that the new epoch would bring different civilizations into closer mutual contact (globalization), with the result that conflicts in the near future would occur at the major 'fault lines' (mind the geophysical metaphor) separating civilizations. In view of the surging violence of Islamic movements versus the West in the current century one is inclined to admit that the thesis was prescient. However, the violence does not manifest itself at boundaries, but rather as terrorist attacks in big cities or in the Islamic world itself. Ultimately, the statistics of armed conflicts have not yet confirmed a significant shift toward wars between civilizations. However, this does not rule out that cultural shock lies at the root of terrorist violence.

A clear indication that there may be more at stake than a 'clash of civilizations' comes from an analysis of Al Qaeda's power of attraction on tribal communities in the Middle East.[13] Most of the hijackers of the planes that crashed into the World Trade Centre towers on September 11, 2001 had a tribal background, a detail that was ignored or even brushed aside in the official report of the 9/11 commission. As Akbar Ahmed has argued, the tribal origin reveals much about the motive of these terrorists. They may all have been Saudi nationals, but as tribesmen they were just as upset by the encroachment of the Saudi national state on their tribal territory as by the Western invasion in the Islamic world. Religion, and particularly Islam, offers the possibility to react to other tribes by claiming genealogical descent from the Prophet Muhammad and by appealing to the quest for religious purity (fundamentalism). From this point of view appealing to Islam is a symbolic weapon in the struggle for territorial purity rather than a clash of civilizations.

The very rise of the great world religions can even be interpreted in this way. At the end of the sixth century the partly nomadic world of tribes in the Arabian Peninsula was threatened by two empires: the Christian East-Roman (Byzantine) Empire in the west and the Persian Sassanid Empire in the east. This encroachment caused alarm among tribes lacking any type of territorial organization, so much so that the phantom of a pair of pincers

---

[13] Akbar Ahmed, 2013, *The Thistle and the Drone*.

(Israel /the West and the Persians) crushing the Arab world was haunting the Arab imagination until the late 20th century.[14]

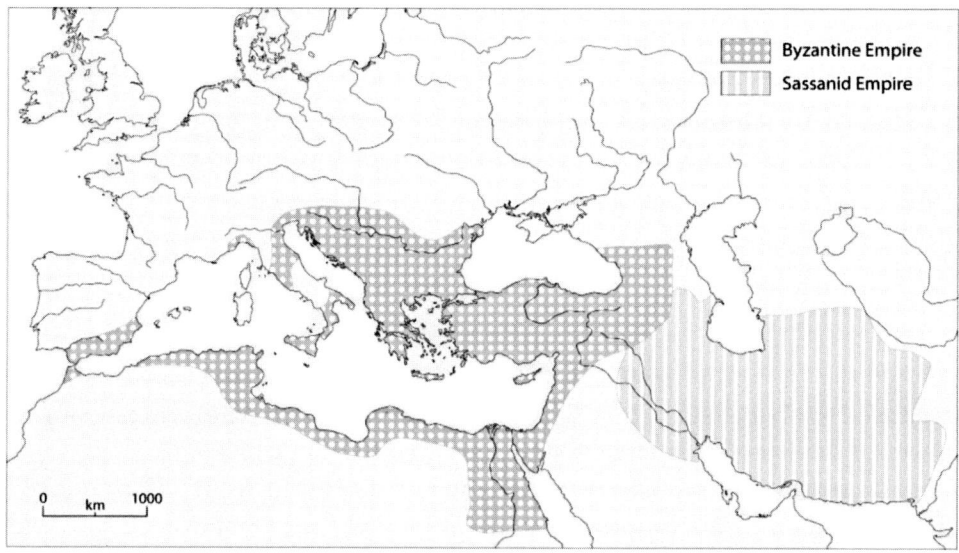

Figure 1.1. The Byzantine Empire and the Sassanid Empire on the eve of the birth of Islam (early 7th century).

There have been several attempts to explain the rise of Islam from socio-economic or geographical factors. One of them was the hypothesis that Islam rose in response to social tensions produced by the booming Meccan trade.[15] Patricia Crone, specialist in early Islamic history, has disputed such explanations by arguing that factors like trade or social tensions are not fundamental enough and not sufficiently specific in time and place (the cradle of Islam was Medina rather than Mecca) to explain such a revolutionary change in people's way of life and outlook.[16] What sort of change and spiritual crisis could have preceded the astonishing march of Islam? According to Crone it was 'imperial trauma'. Only the foreign penetration affected tribal societies that were not able to muster political resources (state institutions) as defence against an alien invasion. These tribal societies responded to the threat with religious

---

[14] Jasim M. Abdulghani, 1984, *Iraq and Iran: The Years of Crisis*.
[15] W. Montgomery Watt, 1953, *Muhammad at Mecca*.
[16] Patricia Crone, 1987, *Meccan Trade and the Rise of Islam*.

means: a monotheistic belief that claimed to match the power of worldly empires by eliminating the multiple and manipulatable gods of the Arab tribes. This religious strategy proved a lucky strike when within the time span of a century the Islamic Umayyad Caliphate ruled over a territory extending from the Indus almost to the Atlantic Ocean.

The rise of Christianity was a similar reaction to the subjugation of the Jews by the Roman Empire. But here we meet with something that was not a call for war but a spiritual renewal aimed at appropriating the changing world order. The problem in ancient Judea was that the territorial dynamics of empire, epitomized by the Roman adage that peace is assured by victories in war and the submission of other people under their authority, did not square with the Jewish perspective that claimed one god favouring one people and one territory. The Romans did not bother about such beliefs and considered Jewish monotheism queer at best, but Jews converting to Christianity embraced such beliefs. They were shocked by the unifying 'transnational' power of the Roman imperial system and tried to reconcile it with their monotheistic tradition, replacing war with love or the 'Holy Spirit'. They created the vision of an all-encompassing kingdom in heaven. The books of the Old Testament never speak about kingdom in this way, only about kingdoms in the Middle East, but the New Testament almost exclusively does. By applying the worldly term kingdom to an all-encompassing authority (in heaven), the New Testament or Christianity reveals a global ambition. It eventually succeeded when the Roman Empire became Christian in the fourth century C.E.

Two years after his death in 44 B.C., the Roman Senate proclaimed Julius Caesar 'God'. His (adoptive) son and successor, Augustus, who ruled when Jesus was a youngster, consequently used the title 'son of God' (*divi filius*). As Showalter remarks, 'Many of those who referred to Jesus as "son of God" knew perfectly well that a Latin form of the phrase was among the most frequent descriptions for Augustus and his successors'.[17] The imperial challenge of Christianity was obvious for its early adherents, but became apparent somewhat later for the rulers in Rome who persecuted Christians. But the Christian religious innovation was an act of reconstruction rather than deconstruction. Crossan and Reed assert: '… both Jesus and [the apostle] Paul are not so much trapped in a negation of global imperialism as establish-

---

[17] Daniel Showalter, 1998, 'Churches in Context: The Jesus Movement in the Roman World'.

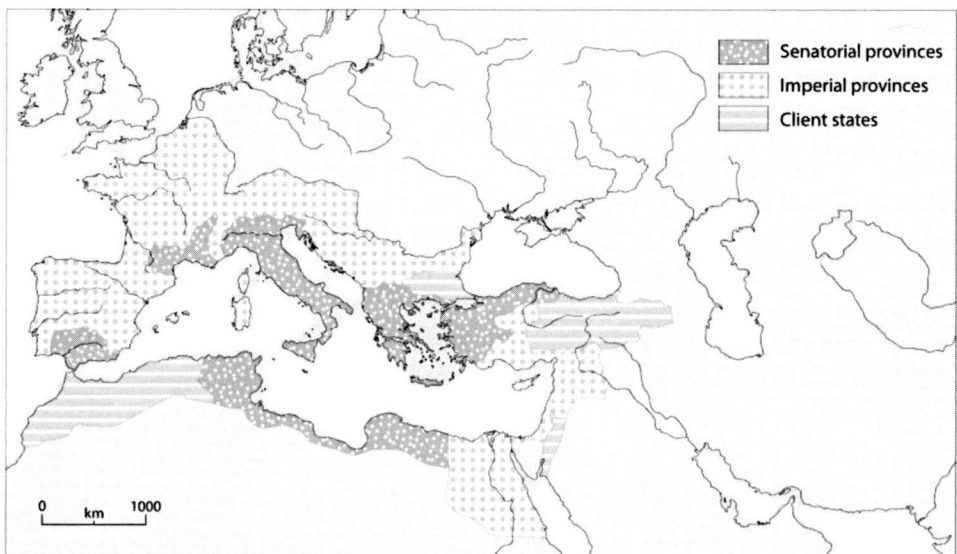

Figure 1.2. The Roman Empire CE 14: Senatorial provinces, Imperial provinces and Client states.

ing its positive alternative here upon earth.'[18] This statement captures the integrating drive that resulted from the shock of experiencing a multicultural polity. However, it was less a clash of civilizations in the current sense of the term than the problem of reconciling the imperial authority with a moral purpose that was equivalent to the Jewish tradition. In this sense the challenge of empire was just as much of a territorial shock as the challenge of the national state by those experiencing the free flow of people and economic activities in our own time.

The Christian assimilation of the imperial territorial order was different from the Islamic reaction to imperial threat that occurred six centuries later. The latter also adopted the alien imperial territorial mode, but from the outset the context made this enterprise more political than the Christian movement. '*Allah's is the kingdom of heaven and earth*' is the typical expression from the Koran, and one should notice the empathic addition 'earth'. Here the hope of building a strong force against an earthly enemy and eventually ruling the whole world is much more pronounced.

---

[18] John Dominic Crossan and Jonathan L. Reed, 2005, *In Search of Paul. How Jesus's Apostle Opposed Rome's Empire with God's Kingdom*, p. 409.

Whether territorial shock involves an external (geopolitical) threat or an 'internal' shift in the way power is exercised and spatially organized, the moral reaction always aims to preserve pride and power by adopting aspects of the new and frightening order. The shock itself can only be explained by understanding the existential foundation of an obsolete order, as indicated by the changes from passive to active in table 1.2. The next chapters will follow the periods indicated in the first column of that table to elucidate the human factor in the territorial history of Europe. But writing history is not the purpose of this book. It is meant to benefit the understanding of our own time as an epoch that is heavily affected by territorial shock expressed in phenomena like populism, new authoritarianism and religious fanatism.

Globalization: fragmented space

Our contemporary world witnesses disturbing transitions: geopolitical, geo-economical and territorial. Among the geopolitical transitions one may count the erosion of the economic and political power of the countries that industrialised in the 19th century, the rising strength of China, and the emergence of other Asian economies. Another change is the susceptibility of the developed world to materially 'lightweight' attacks like terrorism or cyber-war. Democracy and the rule of law seem to have become a liability rather than a strength. The 'war on terror' reveals an aspect of territorial transition: the way in which political spaces have become overlapping and mutually penetrating compared to the mosaic-nature of traditional state-space[19] which ignores all boundary crossing flows in favour of everything that moves within the state space. The need to hunt down terrorists has integrated security agencies in diverse countries, rendering democratic monitoring and decision making difficult. One of the infamous manifestations was the practice of *extraordinary rendition* in which suspected terrorists were transferred to secret prisons in countries outside the US for interrogation and sometimes torturing. One may wonder if such a practice is a symptom of the break in the tradition of warfare, purportedly necessitated by terrorism, or part of a new territorial project that uses the opportunity of terrorism to achieve its aims. According to political advisors and members of the George W. Bush administration, terror-

---

[19] Peter Taylor, 2002, 'Metageographical moments: a geohistorical interpretation of embedded statism and globalization'

ists, by not representing a state, had placed themselves outside the international legal territorial order (the Geneva Convention about the treatment of prisoners of war from 1949) and could therefore be treated at the discretion of any state capturing them.[20]

The 2013 revelations of American security service agent Edward Snowden, which focused on the massive scrutinizing of human information and communication acts (telephone calls, emails, use of internet search machines) by American state agencies, has illuminated another feature of what turns out to be a new territoriality. On the one hand these methods seem to be no more than a perfecting of traditional state control (policing). On the other hand, they facilitate the rise of a transnational security system and transnational manipulation of opinion that are not accessible to democratic control: with unforeseen consequences for *identity*. The misgivings about this development have already penetrated the spy novel literary genre, which is generally committed to the sovereign state and the moral dilemmas that arise while pursuing its security. John le Carré's spy novels never paint a very heroic picture of the intelligence services, but one of his recent novels is even going a step further. In *A Delicate Truth*, the moral dilemma is no longer that secret agents are obliged to commit crimes in the service of the homeland but that they must fight against the authorities of their own country who are involved in transnational business rather than being committed to the common good.[21]

There are less controversial but equally drastic changes in political territoriality, such as the rise of supranational authority in Europe. The willingness of countries to accept the rulings of a European Court of Justice and European Commission was an unprecedented[22] step made by sovereign states. Even if we acknowledge the tensions that supranational governance creates, including unremitting nationalist sentiments, the European Union

---

[20] For example, Alberto Gonzales, Attorney General of the United States from 2001 till 2005.
[21] John le Carré, 2013, *A Delicate Truth*. See also Fintan O' Toole, The Real Men of England. In *The New York Review of Books*, June 6, 2013. TV series like *Homeland* have recently (2018) focused on the clash between intelligence services (FBI) and a government (White House) infected by foreign (Russian) influence.
[22] At least unprecedented if we count out the Holy Roman Empire of the German Nation (1512-1806) which was a remnant of imperial times rather than an innovation introduced by sovereign member states. See for a comparison with the EU Roland Axtmann, 2003, 'State formation and supranationalism in Europe: the case of the Holy Roman Empire of the German Nation'.

has overturned the customary pattern of international relations by adding a higher authority and changing the meaning of boundaries. The new playing field, in which political actors do not exclusively operate on their specifically assigned level but instead involve other territorial actors downward (states using cities as a tool) or upward (states using the EU or another international organization) to reach their goals, has been described as 'jumping scales' or 'glocalization'.[23] Others have dug up the old term 'extra-territoriality' that was introduced in the 17th century to create an exception to sovereign state space for the benefit of diplomatic representation. The principle of diplomatic immunity elicited the practice of recognizing the precincts of embassies as foreign territory. This initially limited form of extra-territoriality is now broadening in scope comprising new forms of waging war, but also establishing special economic zones where different laws apply and controlling the diaspora in other states. This new territorial condition has captured the interest of a varied group of people representing the cultural, philosophic, and academic spheres, resembling a symposium organized in Barcelona in 2005 (Box 1.1)[24]. On the global scale, extra-territoriality was not that exceptional in the period we now leave behind us: the epoch of High-Modernity. It was a commonplace feature of the political space of European colonial powers in the 19th and early 20th century though outside the European continent. Colonialism involved a conception of space that sharply distinguished between the (national) European and the colonial spaces. The application of law differed between the Metropolis (the motherland) and the colony, or rather its natives. A similar distinction held for semi-colonial 19th century China, where European expats were practically above the law. The extension of national law outside the boundaries of the sovereign state bears some similarity to the long-term leases of African land for export-only food or biofuels, which were acquired by Asian countries and international (Western) firms in recent years. Here tracts of land measuring millions of hectares are completely withdrawn from the sovereign state space (and from the food system) of African countries for periods ranging between 30 and 100 years. This is particularly sour

---

[23] Neil Smith, 1996, 'Spaces of vulnerability: the space of flows and the politics of scale'. Erik Swyngedouw, 1992 'The Mammon Quest. "Glocalization", Interspatial Competition and the Monetary Order: The Construction of New Scales'.
[24] Symposium 'Archipelago of Exception', November 10-11, 2005, Barcelona. Text taken from: http://www.cccb.org/en/activities/file/archipelago-of-exception/218418 (Accessed 5-5-2018).

> **Box 1.1. Announcement symposium '*Archipelago of exception*' (Barcelona 2005).**
>
> The symposium will discuss a potentially new cartographic image of the world. It projects the metaphor of the 'archipelago' to help describe a geographical order that is no longer exclusively based upon the model of the homogenous nation-state and continuous borders -- but one fragmented into a multiplicity of extraterritorial zones. These zones are often 'spaces of legal exception' where sovereignty is in question. Reminiscent of extra-territorial zones and the capitulations system of the colonial era where so called 'backward' societies were penetrated by advanced ones – contemporary extraterritorial spaces are embodied today by humanitarian zones, refugee and internment camps, manufacturing enclaves, military bases and some gated communities of nationals abroad. The symposium will describe the different conditions of extraterritoriality and question their political and social implications. It will further reflect upon the challenges that these extraterritorial spaces pose to the modern concept of sovereignty and to the function of human and political rights.

because many such contracts are 'hammered out behind closed doors at the highest levels' without democratic consultation.[25] Another example is the colonization of the 'West bank' by Israeli settlers for whom different laws apply than for the surrounding Palestinian population. The Israeli settlement movement is not merely an economic project. It follows a strategic plan that aims to supervise and intimidate Palestinians and bamboozle international peace negotiators by creating facts that are difficult to grasp in terms of the traditional map as Eyal Weizman has argued. The strategy would particularly be a legacy of general Ariel Sharon.[26]

While extra-territoriality was still an orderly arrangement on the map during the colonial era, it has become much more complicated in a world that is more orderly parcelled out in sovereign territories than ever before. This neat mosaic-pattern, however, is a political Pandora's Box of collapsed states, failing states, not recognized ('pseudo') states, zones with special or limited rights and cross-border connections of state agencies that remain obscure to citizens. The current era seems to witness the demise of the sovereignty principle. The (correct) observation that sovereignty has always been 'organized

---

[25] Joan Baxter, 2010, 'Great African land grab'.
[26] Eyal Weizman, 2007, *Hollow Land. Israel's Architecture of Occupation.*

hypocrisy'[27] (i.e. that sovereignty was never realized in absolute form) cannot be taken as support for the argument that nothing has really changed in the territorial order of the world. Labelling practices such as the African 'land grab' and Israeli settlement policy as 'neo-colonialism' suggests continuity, but today's 'colonialism' is not the result of Great Powers dividing the world amongst themselves. Contemporary globalization is the result of Western capitalism. Further, it made the West dependent on developments in the rest of the world and it also revealed the West's impotence to contain the effects of collapsed or failing states (terrorism, piracy).

Dominique Moïsi has described the basic reaction of the West to the recent geopolitical changes as a culture of fear: 'a sense of loss of control over one's territory, security, and identity – in short one's destiny'.[28] This sounds like an ideal condition for spiritual revival, but Moïsi focuses on a more immediate and visible impact: the disuniting of the West. The disuniting of a geopolitical bloc may be a source of territorial shock, but the analysis in this book rather focuses on the 'loss of control' mentioned by Moïsi. This is the dissolution of the dyad *governance-identity* that constituted the heart of High-Modern territoriality – the possibility to effectively check the dealings of the government and harmonize it with a common national interest. Phenomena like multiple citizenship, the political influence of homeland states on their diasporas and collusion between state governments and multinational enterprises have contributed to this experience of a loss of control and a loss of the meaning of citizenship. Like Moïsi, other writers have alluded to territorial changes in our time and focused on potential political problems and issues for human understanding of the world. Several of them developed a long-term historical panorama. Without pursuing an exhaustive list, I would like to mention some of them to place this book in its academic context.

The past as territorial image of the future

In 1977, political scientist Hedley Bull published *The Anarchical Society*. In this work, he discussed several options for political order in a world in which agencies other than the state have become more powerful than ever. These

---

[27] Stephen D. Krasner, 1999, *Sovereignty: Organized Hypocrisy*.
[28] Dominique Moïsi, 2007, 'The clash of emotions: fear, humiliation, hope and the new world order'.

agencies might include international organizations, multinational corporations, or individual people with enormous fortunes and privileged access to information etc.[29] One possibility briefly mentioned by Bull is 'new medievalism', a condition in which our worldly lords (say states) do not any longer have the absolute sovereignty on everything that happens in a territory but in which new 'popes', warlords and local industries or institutions share or dispute authority over intersecting territories and groups. 'If modern states were to come to share their authority over their citizens, and their ability to command their loyalties, on the one hand with regional and world authorities, and on the other hand with sub-state and sub-national authorities, to such an extent that the concept of sovereignty ceased to be applicable, then a neo-medieval form of universal political order might be said to have emerged'.[30] Whereas Bull's analysis could still be read (in 1977) against the reassuring background of the nested authorities in the European Union, two decades later the term 'anarchy' struck a more alarming note in Robert Kaplan's writings on *The Coming Anarchy* (1994 and later).[31] Here, a journalist travelling through Africa with local means of transport discovered that boundaries do not count for much in the current world. The map is 'a lie' because it suggests an order that is no longer guaranteed, neither today in Africa nor in the near future in other parts of the world. The struggle for scarce resources and between ethnic groups will upset our world order because 'a large number of people on this planet, to whom the comfort and stability of a middle-class life is utterly unknown, find war and a barracks existence a step up rather than a step down'.

Most conceptions of future territoriality oscillate between two positions: the complete erasure of politico-territorial distinctions and the medieval model of overlapping and intersecting authorities. A more sophisticated attempt to paint an image of the future in the spirit of the first category was Michael Hardt and Antonio Negri's *Empire* (2000). In this book the authors suggest that all of us are in the embrace of Empire today.[32] This Empire is

---

[29] Hedley N. Bull, 1977, *The Anarchical Society: A Study of Order in World Politics.*
[30] Ibid., pp. 254-255.
[31] Robert D. Kaplan, 1994, 'The coming anarchy. How scarcity, crime, overpopulation, tribalism, and disease are rapidly destroying the social fabric of our planet'. Kaplan later stroke a different note by converting to the classic geopolitical assumption that geography is the basis of power. Robert D. Kaplan, 2009, 'The revenge of geography'.
[32] Michael Hardt and Antonio Negri, 2000, *Empire.*

built on the global reach of capital and its concomitant pressure toward individual freedom. Although one state (for example the US) may monopolize police functions in order to maintain this new world order, they cannot really capture power because there is no command centre anymore. Power is completely internalized in the actions of countless people and institutions. However, this does not mean that there is equality. People are still subdued on the basis of cultural difference: 'imperial racism'. As genuine post-Marxists, Hardt and Negri applaud the coming of this Empire (like Marx did with the bourgeois state) and they condemn a romantic socialism that appeals to the state as protective shield against capricious transnational capital flows. On the contrary, as Hardt and Negri emphasize, the disappearance of political boundaries creates a new opportunity for 'the multitude' to flex their muscles and enforce a just order. Something of this is perhaps embodied in transnational social movements like Greenpeace, Amnesty International, Attac, etc., but one may with reason doubt their impact.

An early elaboration of the neo-medieval perspective was published by John Ruggie, International Relations scholar and former assistant secretary-general of the United Nations (1997-2001). In a seminal essay from 1993 he characterized the rise of the modern state at the end of the Middle-Ages as the 'consolidation of all parcelized and personalized authority in one public realm' or the 'territorial bundling' of authority.[33] Ruggie compared the bundling of all previously disconnected forms of authority under one central ruler with the rise of the single-point perspective in painting, which occurred more or less in the same period. Just as the details of the new-realistic Renaissance painting are revealed only by supposing a fixed observer, the meaning of each part of a territory can be understood from the point of view of a central agent. The logical sequence of these observations is the conclusion that globalization implies the reverse: the 'unbundling' of authority.

Ruggie did not extend his metaphoric reference of Renaissance painting to contemporary art (virtual space?). However, this frame of mind inspired the geographer John Agnew to distinguish between different ages of geopolitics in Western history as changing visions of the world.[34] His work pertains to visions of the world, epitomized by a criticism of the contem-

---

[33] John Gerard Ruggie, 1993, 'Territoriality and beyond. Problematizing modernity in international relations', quote p. 151
[34] John Agnew, 1998, *Geopolitics: Re-Visioning World Politics*. Agnew particularly aimed at deconstructing the modern vision of the 'parcelized' world as a 'territorial trap.'

porary worldmap as (in Kaplan's terms) a 'lie', rather than to the mechanisms of changing territoriality.

The term 'unbundling', which Ruggie applied to the reversal of territorial authority in our time, finds its economic counterpart in the work of Richard Baldwin.[35] Baldwin describes two essential stages of economic unbundling in the development of 'globalization', which in his conception began in pre-historic times. The first unbundling occurred at the beginning of the 19th century and included the spatial separation of production and consumption which turned some countries into exporters and others into importers. It caused the 'Great Divergence' between the industrialized world and the underdeveloped world (the global 'South'). Since the 1990s another unbundling is in progress, this time due to low communication costs and the development of IT. This unbundling entails the fragmentation of production, meaning some stages of the production cycle, usually the labour-intensive parts, are transferred to lesser-developed countries while other stages like service and innovation remain (for the time being) in the primal industrialized countries. The subsequent rise in national income in the former underdeveloped world – and slackening growth in the US and Europe – creates the 'Great Convergence' and engenders political turmoil about territorial closure.

An author who has focused more on similar mechanisms is Saskia Sassen. She discusses the transformation from *'medieval to global assemblages'* as a change involving three essential elements: territory, authority and rights.[36] Sassen's point of departure is complexity theory, which means that a new territorial order like globalization is an 'emergent property' that can be fully explained from the behaviour of previously existing agencies. More simply phrased: globalization is not an attack on the state but is produced by the states themselves as a new order in which they keep functioning, albeit in a different way. In a similar vein, the French sociologist Jean-François Bayart maintains that 'one of the main lessons of postmodern anthropology will have been to demonstrate how globalization is not so much tolling for the "death of territories" as reinventing them through the effects of "glocalization"'.[37] That notion indicates the coordination between political and organizational processes on differing scales, such as the local, the national and the global. In Bayart's view globalization started in the early 19th century with

---

[35] Richard Baldwin, 2016, *The great Convergence. Information, Technology and the New Globalization*.
[36] Saskia Sassen, 2006, *Territory, Authority, Rights: From Medieval to Global Assemblages*.
[37] Jean-Francois Bayart, 2004, *Global Subjects. A Political Critique of Globalization*, p. 121

European waves of colonization, which actually reinforced the power of national states. However, his main contribution – also in view of our theme of human responses to territorial change – is his emphasis on '*subjectivation*' or the adoption of an attitude and practice that takes advantage of global phenomena but at the same time confers a moral status to the actor. This may indeed be one of the new charismatic religions, like Pentecostalism, which value individual success and enterprise or, to take it further, the imitation of global heroes from sports or movies (African boys who imitate American cowboys).

None of these authors shows much interest in the finer shades of territorial change between 1500 and 2000, although the year 1648 (Treaty of Westphalia) has been venerated in international relations literature as the birth of the 'Westphalian state': the state in which territorial interests eclipsed the dynastic play of dividing or unifying territorial 'properties'. The historian Charles Maier has articulated a more finely articulated scheme of 'territorial regimes' in the period 1500-1980 with three major divisions: imperial (1519-1648), dynastic/territorial (1650-1780) and federal/central (1780-1980).[38] The last stage includes a sub-stage, 'spatial rivalry on the world-scale (1880-1980)', which seems to account for the rise of Geopolitics as an intellectual compass in statecraft . I will not fuss about precise dates or consistency of categories because a more interesting question is **what** we recognise as watersheds between territorial eras. Maier proposes two concepts, 'frontier' and 'field', to outline two major territorial innovations that happened respectively in the mid-17th and end of the 18th century. The 'frontier', in Maier's definition a line separating the territories of two states, ends the existence of 'porous' empires (which were never sure about the loyalty of peoples at the edges of their territory). Unfortunately, the term frontier has been used in other writings, including this book, to indicate precisely this unstable peripheral zone of empires. Conversely, the transformation of territory at the end of the 18th century actually produces a 'field'. The modern state is characterized by a series of links (bureaucratic and communicative) that 'energize' the state and turn it into a single productive body. The metaphors of physical science, in particular the theory of mechanics founded by Newton in the 17th century (geometry, frontiers) and Maxwell's theory of electromagnetic fields in the

---

[38] Charles S. Maier, 2000, 'Transformations of territoriality, 1600-2000'. I have taken the liberty to make the dates somewhat more contiguous and simplify the descriptions.

19th century, seem to neatly fit the changing territorial structure of society.

This birds-eye view of literature about territoriality elicits four types of conclusions:

First, these authors all assume that the study of history provides a key to understanding the contemporary territorial transformation that is popularly known as 'globalization'. This may amount to either searching in the past for some analogy for the coming world order (neo-medievalism), or the idea that studying the transformation of one territorial order into another reveals the mechanisms that produce an 'emergent' reality on the basis of pre-existing agencies and resources. Assuming that history will never repeat itself in the same shape, the latter approach seems more justified, although it has not yielded results that are easy to summarize. Nevertheless, it will be the leading thought in the next chapters.

Second, there is a tendency to relate territoriality to dominant ways of information processing about the world – ways of subjectivation or ways of 'seeing'. The simultaneous rise of the centralized state and of the single-point perspective in pictorial representations, or the analogy between the infrastructural power of the High-Modern state and the theory of electro-magnetic fields in physics, suggest that a territorial order is intricately linked with our orientation in the material world. I will not carry on the storyline of paradigms in physics into the contemporary epoch, but I remind the reader of the remarks about 'orientation' made at the beginning of this chapter. If territoriality is really linked with basic ways of looking at the world, then we may all the more understand that its change induces the 'severe sense of disorientation' that has been reported from areas afflicted with earthquakes.

Third, not all examples of territorial shock presented in this chapter fit easily in the historic perspective on globalization employed by the aforementioned authors. Christianity and Islam were primarily reactions to submission to a foreign power or to the rise of empires. Although reaction to submission or to the threat by a foreign power is not a very good description of the central interest of historians studying re-territorialization (in Europe), I have applied the label 'territorial shock' because in both cases, Christianity and Islam, the threat came from an alien territorial order. Moreover, these cases are interesting because they show how spiritual defence against a territorial threat creates a *modus vivendi* that is more than resignation; it is empowerment. Or in

Bayart's terms: creating a moral subject. In both cases the reaction foreshadowed re-territorialization because it created an Islamic empire in one case and people who could move in a reinterpreted 'global' Roman Empire in the other case.

Finally, the theoretical statements on territoriality discussed above deal silently with only the highest scale or the strongest political authority, that of the state, although we know that there are other levels of political territoriality (regions, cities) which might have an emotional impact on people when subjected to change. This was particularly valid for medieval cities, a kind of proto-state (see Chapter 3), and their struggle to remain independent. Regions may become stronger or weaker over the course of time (more closed or open, institutionalized or eroded by the state, emotionally charged or a mere formality) but this is often a corollary of state territoriality – not in terms of a zero-sum game (more state = less region), but as an intricate process of 'glocalization'. This is often implied in the observations made on changing territoriality in the next chapters.

These conclusions point to an encouragement to study profound territorial change as a subtle interplay between material and spiritual change. A new territorial order implies a new way of looking at the world, including a new moral principle that makes the world acceptable. However, most studies mentioned above do little to analyse the painful areas in the process of reterritorialization (those producing territorial shock) and the moral response they demand. This is the aim of the present book, which of course could never have been written without the intellectual achievements of these authors.

Chapter 2
BARBARIANS AT THE GATES: THE CLASSIC EMPIRES

> A bright moon rising above Tian Shan Mountain,
> Lost in a vast ocean of clouds.
> The long wind, across thousands upon thousands of miles,
> Blows past the Jade-gate Pass.
> The army of Han has gone down the Baiteng Road,
> As the barbarian hordes probe at Qinghai Bay.
> It is known that from the battlefield
> Few ever live to return.
> Men at Garrison look on the border scene,
> Home thoughts deepen sorrow on their faces.
> In the towered chambers tonight,
> Ceaseless are the women's sighs.
>
> **Li Bai** (Li Tai Po) 701-762

> This is the order of the everlasting God In heaven, there is only one eternal God; on earth there is only one lord, Chinggis Khan. This is the word of the son of God, which is addressed to you.
> *Preamble to a letter to Louis IX of France by the Mongolian ruler* **Möngke Khan** 1255 [1]

The gates of Empire and other semantic problems.

War is the most ancient source of emotions evoked by boundaries. But the 'women's sighs' and 'home thoughts' of Li Bai's poem are not the expression of territorial shock; they simply witness the inevitable loss of life at the border. This is the world of empires: where the frontier is an unstable zone keeping barbarians at a distance, by constant vigilance and the exercise of violent revenge. An empire's boundary does not delimit a special 'race'; its social cohesion is a matter of allegiance to the Emperor or God. Everything is organized in a vertical way, while horizontal links are limited and

---

[1] Eric Voegelin, 1940-41, 'The Mongol Orders of Submission to European Powers, 1245–1255', p. 391.

discontinuous. It is striking how our language has held on to this special organizational feature that is markedly different from that of the modern state. Our language allows us to associate the wildest of territorial images with the idea of Empire – The Empire of Dreams, of Evil, of Goods, an Island Empire, the British Empire and the Empire of Lights. All these linguistic forms suggest a collection of things or actions ruled by some principle, but not necessarily within a clearly demarcated and contiguous space.

Empire des Lumières, the title of a project by painter René Magritte, pushes even further by extending the principle of non-contiguity over space *and* time. The paintings, a series of several works produced over a prolonged period, usually show a nightly landscape with houses and above it a sky in broad daylight. But there are also islands of light down in the darkness: illuminated windows and streetlamps. The Empire of Lights covers events that happen simultaneously but never in the same spatial setting, akin to the British Empire, 'on which the sun never sets'.

In many of these cases the English language accepts the word 'kingdom' or even 'republic'[2] instead of empire, but the word 'state' appears inconvenient. The semantics of 'Empire' hint at an unbounded sphere and an ambulant or heterogeneous power that strikes to inspire awe. 'State of fantasy' would be an inconceivable expression, although one cannot deny that the state has succeeded in achieving one of the greatest fantasies in human history: national identity.[3] Yet the state seems too mundane, too practical, for the phenomena invoked by the concept of 'Empire'. One of the main achievements of the modern state is that its territory has been mapped, or to put it in a different way, made 'legible'.[4] This mapping entails much more than demarcating borders; it is an appropriation of territory from a specific point of view. This may include the introduction of uniform measures and units, taxation standards and the complete partition of the territory by a hierarchical system of administrative units. The state requires fixed positions for all its moving parts, which over time become increasingly interconnected by

---

[2] For example, 'the kingdom of animals', 'the republic of letters'. The Germanic words for kingdom still reveal a link with empire in their word-syntax: *Königreich*, *Koninkrijk* (=King-empire). Other words in the English language are realm, commonwealth, and territories.
[3] 'Fear' would be acceptable in both the context of state and Empire. Fear apparently flourishes in strongly and weakly organised settings. 'Republic of Fear' is a book by Kanan Makiya (1980) about Saddam Hussayn's regime.
[4] James C. Scott, 1989, *Seeing like a state*.

roads, pipes, and wires. No wonder that the first treatise on political geography, Friedrich Ratzel's *Politische Geographie* (1897)[5], found the essence of the state in its attachment to the land (*Boden*) and its organic appearance. The state's physical nature tolerates neither territorial discontinuity nor uncontrolled spaces. Where empires are foremost based on personal accountabilities and words of faith, states surround the individual with material equipment and a disinterested bureaucracy, which guides individual behaviour on a (so to speak) 'molecular' level. This principle of control is one of the basics of political territoriality as defined in this book.

Undoubtedly, the material-based modern state is hard-pressed to survive in a period that is characterized by wireless connections, air transport, mass migrations, cosmopolitanism and symbols that have the power to mobilize masses around the world (Islamic fundamentalism, transnational identity). Technical inventions like television and Internet seem to resurrect the discontinuity of empire, and with it the possibility of spiritual unity over a large heterogeneous area. Hedley Bull's (1977) description of a future world order as 'new medievalism' is surprisingly appropriate, particularly in Europe where states have ceded or abolished so much of their autonomy in the name of a weak political body (the EU) that time and again faith is demanded to accept its rule. One may recall the history of the Middle Ages in which the Empires of Charlemagne and their German successors endured thanks to their moral appeal on the local lords rather than the ability to run a smooth political and military machine. No doubt we may think up many differences between the current situation and the Middle Ages, but the comparison at least draws our attention to the fact that in our life the comprehensiveness of state territory is no longer obvious. Moreover, it tells us that although the history of the modern state is very short, it has brought us advantages that we cannot easily do without.

If the semantics of 'Empire' suggest an order of things with an elusive materiality, this does not do justice to all aspects of classical Empires – like Charlemagne's, the Roman, the Ottoman and the Chinese Empires. Most of these were struggling to behave like states, at least in a territorial way, and are therefore often rewarded with the title 'state' by historians. It means that emperors tried to consolidate their authority over a continuous area and that they were reinforcing boundaries against invasions of barbarians. Yet this was

---

[5] Friedrich Ratzel, 1897, *Politische Geographie oder die Geographie der Staaten, des Verkehres und des Krieges*.

only partly successful. Pockets of resistance persevered in the Empire: a feature that the authors of a French comic strip about the intransigence of a small Gallic village against the Roman conquerors have exploited as a rich source of entertainment.[6] The high and wild place of Montenegro remained uncontrolled during centuries of Ottoman dominance in the surrounding areas, even when there was more effective imperial authority in the valleys. Empires failed to treat everybody in the same way. They had to accept local strengths and powers; if necessary, tax collectors and recruiters went out of their way to avoid them. In the cultural sphere, even the will to equalize was absent in empires. In pre-industrial society, groups with a different culture often practised special activities such as loaning money (Jews in the Carolingian empire) or handicrafts (non-Arabs in the Umayyad Empire). These forms of segregation only advanced political stability since they discouraged attempts to get involved in 'alien' activities like administration or the wielding of power. Everyone had and knew his or her pre-ordained place in this universe.

That birds of different feathers all recognized the same emperor only affirmed his status as true ruler over the world. From the 17th century on, Russian Tsars, although head of the Orthodox Church, resisted attempts to Christianize their Tatar (Muslim) subjects.[7] This tolerance, however, had not always been practised in previous times of war. When Tsar Ivan the Terrible expanded the frontier of Russia eastward against the Kazan Khanate, Tatars were forced to convert to Christianity. Otto I (912-973) did the same with the western Slavs in the 10th century. In some cases Turkish conquerors gave their defeated opponents the choice between death and conversion to Islam. But in times of political stability, communities of Christians and Jews within the Ottoman Empire (1326-1922) were not persecuted for their religious beliefs (although the full rights that Muslims enjoyed were not granted to them). In light of recent experiences with ethnic hatred and genocide, some authors have been led to positively revalue the politics of empires such as the one that for so long ruled over the Balkans.[8]

---

[6] Uderzo and Goscinny's comic 'Asterix'.
[7] Andreas Kappeler, 1993, 'Some remarks on Russian national identities (sixteenth to nineteenth centuries)', p. 150. See also Mark Bassin, 1999, *Imperial Visions: Nationalist Imagination and Geopolitical Expansion in the Russian Far East*.
[8] Daniel Goffmann, 2002, *The Ottoman Empire and Early Modern Europe*. Similar revaluations have been made about the Habsburg Empire.

Chapter 2. BARBARIANS AT THE GATES

Empires appreciate sharp boundary lines, like states do, but the difference is that they never really succeed in getting those boundaries to work as reliable dividers of power realms.[9] The basic problem is the absence of a symmetrical (bilateral) relationship between political systems on either side of the line (since an Empire considers itself as 'the civilized world' and the rest as nobodies, barbarians). While empires show no real will to abolish internal cultural differences, they neither want to abandon the idea that people outside the political heartland should be subservient to the empire. The result is that the periphery of an empire develops into a buffer zone, a *frontier* rather than a region where people are divided by a boundary line. Frontiers require frequent military expeditions that do not necessarily expand the heartland because resources or ecological conditions discourage the establishment of civilized life in that area.[10] A typical outcome of this condition is the formation of an intermediate culture, born by people with a way of life that does not meet civilized standards, who even proudly ignore the heartland and its values, but nevertheless act as a defence against the barbarians. This is the classic image of the American frontier: the Wild West recruiting its inhabitants from outcasts of the 'Metropolis' who gradually tamed the Indian wilderness. It is the clearest example of the dictum that people in frontier areas turn their backs on the centre whereas people near (state) boundaries turn their backs on the boundary.[11] This also explains why Chinese attempts during the Han (206 BC-AD 220) and Ming (1389-1644) periods to solve logistic problems by having soldiers settle as farmers at the boundary failed. They could perhaps grow their own food, but like the soldiers in Li Bai's poem they were troubled by 'home thoughts', even if these soldiers were convicts.[12] However, attempts to organize a frontier by means of settlements seem to have gained some success in Europe. The confrontation between Europe and the expanding Ottoman Empire between the 14th and 17th centuries displayed all features of the entanglement of a civilization with an 'inferior' outside world. From the

---

[9] One may wonder how much state boundaries are reliable, particularly from the time perspective of the classic Empires, but the entire political attitude towards them is completely different in states.

[10] Ratzel already referred to the necessity to adopt the way of life of the adversary if one wants to conquer them. Friedrich Ratzel 1897, p. 80.

[11] Ladis K. D. Kristof, 1959, 'The nature of frontiers and boundaries'.

[12] Arthur Waldron, 1990, *The Great Wall of China. From History to Myth*, p. 42. Thomas J. Barfield, 1989, *The Perilous Frontier. Nomadic Empires and China*, p. 54.

16th century on most of the defensive burden fell on the Habsburg Empire. Its armies made deep incursions into the Turkish area but again and again they had to retreat to the demarcation line of the Sava and Danube rivers (the current boundary between Croatia and Bosnia).

Figure 2.1. The military frontier (Vojna Krajina) in the Habsburg Empire. Source: C. & B. Jelavich, 1977.

In their wake followed Serbian refugees who sought political asylum in the Habsburg Empire. The authorities decided to allot these migrants land at the frontier and provide them with weapons. This 'military frontier' or *Vojna Krajina*, a zone at least 25 kilometres wide at most areas (figure 2.1), became a refuge as it was withdrawn from the authority of local Croat lords and placed under immediate supervision of the Habsburg Emperor in Vienna.[13] Here political and cultural conditions truly created a society of which the members could look outward and forward (to the future liberation of Serbia). In the 20th century, however, these Krajina Serbians living in the Croatian area would become part of the ethnic tangle that shocked Europe so much during the dissolution of Yugoslavia.

---

[13] Stevan, K. Pavlowitch, 1999, *A History of the Balkans 1804-1945*. Charles and Barbara Jelavich, 1977, *The Establishment of the Balkan National States, 1804-1920*. See also the discussion on the medieval frontier further down in this chapter.

The Walled Empire

If buffer zones are inevitable in securing an empire's territorial integrity, why then did the Chinese hide behind a huge wall?[14] The Great Wall is a textbook case of a defensive shield erected by an agricultural society against nomadic tribes with unfitting ways of life. Is it not likely that the Chinese still operated a wide zone for military operations beyond the wall? The fact that the current northern boundary line does not correspond with the course of the Wall indicates that movements must have occurred across the line. This movement could have been a recent event, dating from a time when military technology had reached a destructive capability with which no great wall could compete. There are, however, indications that the Great Wall never stopped any purposeful movement and that it was actually only recently (early 17th century) constructed, or rather finished. How to explain this paradox? What to think of the geometrically correct Chinese maps from the 13th century (as reproduced in Needham's classic *Science and Civilization in China*) that so unambiguously trace out the Great Wall? And what about the many scholarly works that mention the Wall as 'an enormously ambitious construction originally completed by the Qin dynasty at the end of the third century BC to physically demarcate and separate China from the world of the nomads'?[15]

Arthur Waldron has argued that the annals of the Qin (221-206 BC)[16] make mention of armies leaving outposts or crossing mountains but not of walls being passed. The Mongol invasions of the 13th century left no reports describing how a wall slowed their advance, nor did Marco Polo mention a Great Wall. Moreover, there is no word or symbol in classical Chinese that can be interpreted as reference to 'the Wall'. This seems to be contradicted by early Chinese maps that at least present the image of a single long wall. Waldron states that the maps are not realistic since the Wall passes almost entirely through territory that the Qin dynasty never controlled. Therefore, it is "perhaps drawn not on the basis of a survey, but rather according to written

---

[14] Similar questions have been raised about the boundaries of the Roman Empire which have become widely known to the public by the excavation of fortifications, watch-towers and walls (Hadrianus Wall in the north of England). As Stephen Dyson remarks: '…walls and forts were only part of a larger diplomatic, military, political, social and economic system that embraced both sides of the frontier and created a gradual transition from Roman to non-Roman society'. Stephen L. Dyson, 1985, *The Creation of the Roman Frontier*, p. 3.
[15] Barfield, 1989.
[16] The Qin became generally known for the archaeological discovery of the terracotta army.

records and traditions". This may be true, but the argument does not extend further to the simple fact that a long wall at least existed in the Chinese *imagination*. Anyhow, written sources do not appear to give evidence of a single wall. If texts refer to the frontier it is with phrases like 'the nine border garrisons' (*chiu-pien-chen*), and if fortifications are mentioned they are denoted as 'border walls' in the plural. One can only conclude that the Chinese, during the major part of the last two millennia, attributed the construction of a Great Wall to the Qin empire, but in the majority of military reports a wall was never encountered. Indeed, Waldron maintains that classical texts about the general territorial aspects of the Qin refer to 'fortifications' rather than to a Great Wall.

Still, there is no question about the significance of walls in early Chinese history. Walls surrounded the Chinese states that preceded the first unified Chinese Empire of the Qin in 221 BC. As the explorer and self-made geographer Owen Lattimore wrote in 1937:

> (…) the earlier systems of frontier walls that preceded 'the' Great Wall cannot be attributed entirely to the wars between Chinese and 'barbarians'. Among the earliest of them were walls that ran north and south, between one Chinese state and another…There must have been something inherent in the historical process of the state in China that favored the evolution of walled frontiers, irrespective of hostile reactions between the Chinese and peoples whose ways of life where incompatible with theirs (…)[17].

This background suggests that wall fragments probably preceded the current Great Wall and since such walls were ubiquitous, no particular attention could be expected in places where they marked the border with nomadic tribes or 'states'. Walls were generally thought fit for separating areas that were culturally or ethnically distinct, in a similar way as Heaven had sorted out natural regions. Early Chinese strategists realized that walls alone could never solve the threat of nomads like the Xiong-nu. Action beyond the borderline was deemed desirable, first because it fit the universalistic ideal of Chinese power and second because it allowed the use of more effective means to avert external dangers (for example by spreading discord among the nomads, building exchange relationships or offering help and implementing pre-emptive mili-

---

[17] Owen Lattimore, 1959, 'Origins of the Great Wall in China: a frontier concept in theory and practice', p. 58.

tary strikes). All such strategies require a self-assured empire that allows itself to enter into peaceful dealings with the outside world because it knows that, as a last resort, it can revert to military power. Obviously, such conditions cannot continue forever. Within each dynastic era periods of weakness occurred, revealing tendencies to withdraw behind boundaries and to cut off relations with the outside world. This introverted attitude went hand in hand with an ideology that emphasized Chinese ethnic identity, contrasting the 'dynastic' approach which big-heartedly defined the empire as a collection of people loyal to the emperor (a political choice with a familiar echo). An introverted period produced (or completed) the Great Wall, and it was **a sign of weakness rather than strength**.

The Ming (1368-1644) started in an extroverted mood. Its founder, the Hung-wu emperor, established the 'eight outer garrisons' aimed at controlling the Mongols deep into the steppe. More inwardly looking, he erected a line of fortifications, a forerunner (as Waldron would call it) of the Great Wall. Furthermore, he introduced a system of military farming that was supposed to feed an enormous army without burdening the peasants. None of these policies were granted a long life. Military farming could not work and under later emperors the outlying garrisons were drawn inland one after another. In due course the mindset of the Ming Chinese changed. New intellectual currents emphasized the cultural identity of China, and the practical geopolitics of the frontier gave way to an idealized conception of Chinese superiority that had to be confirmed in the *tribute ritual*, the ceremonial offering of gifts by foreigners to the Emperor to show their subservience. Yet the choice between extroverted and introverted policies still evoked controversy because an assertive attitude towards the outside world was still seen as a matter of honour. Such debates and intrigues fueled the political and intellectual crisis of the late Ming, which produced an exceptional number of reflexive documents and even literary dramas on political issues. These documents are highly interesting for our understanding of Ming society, but in their time they only reinforced the inward turning spiral of politics and the decaying agility in foreign policy making. The building (or consolidation) of the Great Wall was a logical outcome, but it can also be interpreted as a symbol of the Ming dynasty's inability to act upon the outside world, eventually sealing its fate. The fresh wall proved to be no obstacle for the Mongol hordes that soon overran China, creating the Manchu (Qing) dynasty (in power until 1912). China, as an empire that emphatically considered itself the centre of the

Figure 2.2. Foreign delegation with tribute (fragment). Yen Li-Pen (Tang 618-907).

world, Kingdom of the Middle, ruled by the Son of Heaven, exemplifies the practical difficulties that can arise while dealing with the rest of the world. Ritual performed at the court had to be seen from the perspective of this cosmic order. 'In the annual fertility rite [the Emperor] plowed the sacred furrow not so that Chinese crops would grow but so that crops per se could grow'.[18] Foreign merchants could not be acknowledged as providers of anything that China lacked, even if everyone knew better. Their activities were regarded solely as the visit of a diplomatic delegation that came to pay tribute to the Emperor. At the end of the 19th century, European commentators still felt irritated about the failure of China's ministers and people to 'realize the fact that China is only one of many independent nations in the world'.[19] Again China was blind to the fact that an external political reality, European colonial expansion, had already wiped away the few elements of its world order that had ever had some reality.

---

[18] Mark Mancall, 1968, 'The Qing tribute system: an interpretive essay', p. 64.
[19] James Legge quoted in Lien-sheng Yang, 1968, 'Historical notes on the Chinese world order', p. 22.

Any empire must navigate the contradiction between their own universalistic aims and the reality that they are impotent to control events outside the empire's Metropolis or core (*'imperial overstretch'*). Appearances can be kept up for some time by symbolic actions or by withdrawing into a closed world of words and ritual, but the latter only hastens decline. This happened repeatedly in Chinese history. The nomadic fringe was the most significant threat to security and the most likely cause of death to dynasties. Yet, at the same time, the nomadic fringe was the target for policies aimed at mellowing these people while simultaneously serving the Chinese self-perception of superiority. The involvement of the Chinese with the outside world was explained by the Emperor's wish 'that none of the ten thousand countries in distant lands should not be his subject'.[20] Actually, the contact between the Chinese and nomadic groups was initiated by the nomadic groups as much as by the Chinese empire. This was because the nomadic way of life could only flourish in a symbiotic relationship with a sedentary society.[21] Nomadic rulers were eager to enter into relationships which the Chinese interpreted as tribute, but the real reason behind the rulers' outreach was a desire for valuable goods such as wine, corn, silk and metal utensils. Once, a 'delegation' to the Emperor numbered two thousand, and they all counted upon generous recompenses. From time to time Chinese political commentators couldn't take it anymore and insisted upon radical measures. In Han times *Jia Yi* (201-160 BC) complained about hanging upside down:

> The situation in the empire may be described as like that of a person hanging upside down. The Son of Heaven is the head of the empire. Why? Because he should remain on the top. The barbarians are at the feet of the empire. Why? Because they should be placed at the bottom. Now, the Hsiong-nu [Xiong-nu] are insolent and arrogant at the one hand, and invade and plunder us on the other hand, which must be considered as an expression of extreme disrespect toward us. And the harm they have been doing to the empire is boundless. To command the barbarian is the power vested in the Emperor on the top, and to present tribute to the son of Heaven is a ritual to be performed by the vassals at the bottom. Now the feet are put on the top and the head at

---

[20] According to the Yung-lo Emperor around 1400. Joseph F. Fletcher, 1968, 'China and Central Asia, 1368-1884', p. 207.
[21] Barfield, 1989.

the bottom. Hanging upside down like this is something beyond comprehension.[22]

Complaints like these were a thinly veiled recommendation for the Empire to make its powers felt, but later Han historians realized that it could have been a shrewd strategy to put the Xiong-nu in their proper (subservient) place to be prepared for a future when the Empire was weak. Pan Ku (CE 32-92) warned against too much interaction with the Xiong-nu, arguing that it could draw them too far into the Chinese sphere. He advised treating them as guests that one could never completely trust. This introduced a distinction between barbarians in an 'inner' and 'outer' sphere. Outer-barbarians should not demand any formal relation, neither should they be targeted by military operations unless they attacked China.

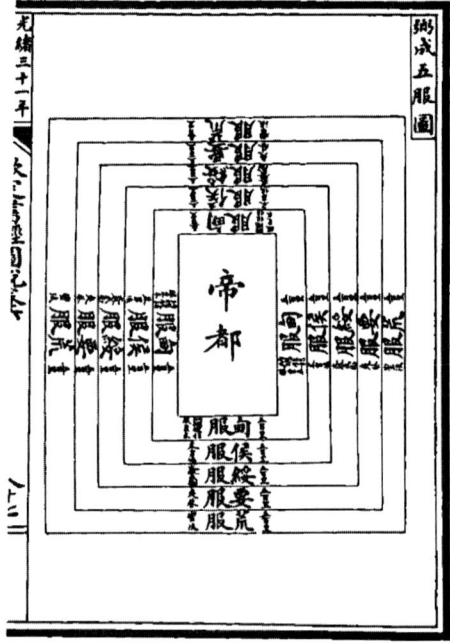

Figure 2.3. Model of China and its environment from the Yü Kung chapter of the Shu Ching (Historical Classic), ca 500 BC. Reproduced from Needham 1959 vol. 3, p. 502.

---

[22] Ibid., p. 53.

This soul-searching about the kinds of people and places that should be distinguished in terms of strategies and rituals is more characteristic of classic China than its attempt to close off the outer world by means of the most drastic actions like the building of a Great Wall.[23] The oldest representations from the 5th century BC show China as the centre of a number of nested squares that represent different domains requiring different rituals or responses (figure 2.3). From the centre to the periphery these were: a. the royal domains, b. the lands of the tributary feudal princes and lords, c. the zone of pacification, d. the zone of allied barbarians and e. the zone of cultureless savagery.[24] In the Chinese ontology the world is a system of natural forces and fields that differentiate places and in this way imply boundaries.[25] Chinese landscape painting suggests an infinity from which mountains and rivers emerge: quite different from the Western representation that shows the world as a container that can be filled with different types of objects (however, it was only after the Middle Ages that this 'Western' form of representation emerged).

The Holy Empires

The Medieval Roman Empire

During the greater part of the Middle Ages (roughly from 400-1400), Empire-Europe was an ideal rather than a reality. It owed its attractiveness to the binding power of Christian religion rather than to some kind of effective administration or loyalty to an emperor. Such emperors as did emerge – legitimated and anointed by the pope – could never claim to rule over entire Europe or even a substantial part of it, but they match what we have already established about emperors: there *claims* of sovereignty were always unbounded. They considered themselves as the worldly caretaker of God's Empire, which, unlike the analogous Chinese case, elicited jurisdictional struggles with a pope. The medieval world was a universe of personal relationships in which territoriality in a modern political sense did not exist. The one who could gather the courage, weapons, and the tactical vision to prevent others from occupying a piece of land could call himself a sovereign power, usually

---

[23] see also the quote of Lattimore on p. 34.
[24] Joseph Needham, 1959, *Science and Civilisation in China. Vol. 3*, p. 502.
[25] John Hay (ed.), 1995, *Boundaries in China*.

sealed with the impressive landmark of a fortified residence or farm. When conditions turned more chaotic and dangerous from the 9th century on, some freemen sought the protection of more powerful freemen, becoming a *vassal* and part of the protector's army. Conversely, kings and barons could grant land in *fief* in exchange for military service. At the bottom of this hierarchy in which vassals could create vassals of their own stood serfs, who were granted the use of the land in return for personal services. Serfs were not slaves but in practice they had little freedom to resign from their job (the penalty being condemned as an outcast without means of existence). This is, briefly, the notorious feudal system that took shape in the second half of the Middle Ages. The label 'feudal' has in recent times been attached to all kinds of systems of social exploitation, but in the absence of a central authority feudalism undoubtedly fulfilled a useful security function during a long period in the Middle Ages.

A vassal's land may have been granted in fief from the part of a king or baron, but this did not imply that the king or baron had any authority over vassals or their serfs. It only implied some obligations to the lord. Next to such personal juridical relationships there were free cities, dioceses with their own land and administration of justice and villages with inalienable customrights on the use of surrounding land for grazing, picking berries or the gathering of firewood. People knew very well how to demarcate land for the benefit of rights and utilization, but there was no concept of territory as a completely closed system of authority, let alone of public representation.

A hypothetical medieval traveller through Europe would not experience the transgression of boundaries in the sense of entering a new territory with a different rule of law or with surveillance agencies that could interrogate or arrest 'foreigners' coming from elsewhere. Of course, our traveller would meet different customs and languages and would have to pay tolls at the most unwelcome moments. He could even fall victim to a malicious *seigneur* or to highwaymen, but essentially his passing through space would not put his identity to the test, which firmly remained fixed within a network of personal relationships and the all-encompassing Christian world order with bishops ruling as worldly lords and Latin as its *lingua franca*. The heterogeneity of 'political space' – the network of rights and obligations impinging on the person – would not be different from that with which he was familiar in his home region. There the king or emperor may have had rights on the wilderness, on hunting or fishing and on administering justice over the freemen, but again not over the serfs of local lords or even over the forests within their territories, neither over the clergy or cities. Territory was not a good cue for knowing whose justice and administration applied because so many overlapping places were exempted from the power of authorities, even those classified as 'high' authorities, because the latter ruled over a wide area or had many vassals.

The Roman Empire was the predecessor of the medieval European empires, but its continuance owed more to the Christian spiritual heritage of the late Roman Empire than to the persistence of territorial power. Before the adoption of Christianity as its imperial ideology, the Roman Empire was rather what Peter Brown has aptly described as a 'commonwealth of cities' in which upper-classes, entitled to tax collection, were the actual agents of the Empire.[26] There was no centralized control (from Rome) on the ins and outs of local societies but rather a 'horizontal…collaboration of an empire wide upper class drawn from the elites of the cities'.[27] The main interest of the imperial court was in maintaining military strength rather than regulating society or 'integrating' (to use a modern term) its subjects. In the 3^d century C.E., however, military projects and internal coordination started to fail. The Empire was in danger and responded by making its presence more keenly felt in local society by demanding loyalty and accountability. This diminished the role of the local upper class as representative of imperial government. In this climate with its new urge for unification, Christianity fitted as the ideal moral and philosophical underpinning. It became the official doctrine after the new Emperor Constantine attributed his success in war (312 C.E.) to the blessing of the High God of the Christians.

What was so special about Christianity that it could help to amend the failings of the old Roman Empire? One attribute resonating with the new centralizing ambition was the Christian idea that there was one God and one divine Law that applied to everyone. The old Roman religion blended smoothly into a world in which local and particularistic interests meshed with the belief in a diversity of gods and spirits that could be invoked to serve those interests. Conversely, the universalistic claim of the Christian belief fitted the social composition of the community that congregated in the churches. This was a mixed community consisting of high and low classes,

---

[26] Peter Brown, 2013, *The Rise of Western Christendom: Triumph and Diversity, A.D. 200-1000*, pp. 54-56.
[27] Ibid., p. 57.

men and women and people representing different crafts: briefly, the demography of the Empire. Further, the Christian emphasis on self-transformation, cleaning people from sin by conversion, baptism and penalizing wrongdoing resounded well with the idea of an empire that could renew itself, not by co-opting new elites but by transforming itself from the bottom up by infusing itself with a new spirit. This was wonderfully supported by an institutional structure of bishops and parishes, which had proved their capability to weld different classes together and to provide a form of secular governance as well.

The material presence of the Roman Empire in terms of military force, monumental buildings, cities, and infrastructural works in Western Europe faded away in the course of the 5th century as a result of weakness of the Centre (Rome). This left an 'orphaned' Roman aristocracy and a new class of military men, today one would say 'warlords', who were portrayed by a contemporary source as barbarians in deer-skins. They were not, however, the notorious invading barbarian tribes, but groups that had earlier migrated to the Roman Empire, lured in by the Roman need for military forces. Peter Brown opines that talk about '*the age of barbarian invasions*' constitutes a misleading picture that had already originated with the Roman eagerness to describe people across the *limes* (the frontier) as barbarian.[28] It is a well-known way of making boundaries ex-post-facto logical.[29] Actually, there was no great difference between the people within the imperial territory and those without, they were all agrarians (unlike the situation at the Northern frontier of China). But these fur-clad militia men, who locally started to build territories in response to the need for law and order, had often been in imperial service for generations.

From one of these groups descended the Frankish kings that enlarged their territorial power until their leadership devolved upon one man: Charles, the later Charlemagne. Pope Leo III manoeuvred in such a way that the Church seemed to wield the power of the Roman Empire and to be authorized to transfer it to a specific emperor. The legendary coronation of Charlemagne as Emperor (800) actually started an age-long juridical struggle about the authority of the pope in worldly matters and (somewhat less) of the Emperor in spiritual matters. The consequences were far-reaching. First, when the Empire was falling apart or had to acquiesce to the growing autonomous

---

[28] Ibid, pp. 101-106.
[29] See also the representation of a man-made boundary as a mountain range in the history of the Pyrenees Peace Treaty, Chapter 3 (figure 3.3).

power of specific kingdoms, the emperor showed a tendency to take over authority on religious affairs (England) or to assume a special link with God (France). Second, when the imperial title was transferred to the German kings (start of the Holy Roman Empire with the crowning of Otto I in 962) the Empire had to face regional lords or princes that already insisted upon a 'natural' right to rule their territories without external interference. Although not remotely democratic, they nevertheless felt that their effectiveness as rulers would benefit from a degree of local legitimacy rather than from authority that was ultimately derived from an Emperor sanctioned by God. The repercussions of this dynamic on the nature of the Christian religion will be discussed in the next chapter.

How did the Empire look from a geographic point of view? Weak political boundaries went along with frontiers, where the established medieval order turned into heterogeneous and unstable societies that separated the Christian Empire from the distant lands of the 'barbarians' which, by the way, included civilized worlds such as the Muslim and Byzantine empires. The under-populated Polish-German frontier, although Christianized from the 10th century on, was for a long time the scene of settlers coming from the west. The Polish rulers welcomed them because they needed troops (like the knights of the Teutonic Order) in their expeditions against the pagan tribes in East Prussia or the Kiev state in the east, but also because foreigners introduced new agrarian methods. Besides, Poland had already been the scene of involuntary settlement of individuals captured in wartime.[30] This creates a picture of an environment prone to local rivalries and instability. In other European peripheries, for example the frontier separating Europe from the Caliphate of Cordoba, settlement was more directly organized for the purpose of warfare. In Spanish Castilia the *fuero* (by-laws) of Sepúlveda allowed the town to become 'a refuge for assassins, adventurers and outlaws' taking advantage of the privilege of *homicianos* (1076) or amnesty for criminals.[31] The mere fact that someone settled in the town was sufficient to drop any existing charges against that person. Other settlers, free men without criminal record, could easily get access to the group of knights, provided that they took along a horse. Such suspension of established norms and class differences is a

---

[30] Paul Knoll, 1989, 'Economic and political institutions on the Polish-German frontier in the Middle Ages: action, reaction, interaction', pp. 151-174.
[31] Manuel González Jiménez, 1989, 'Frontier and settlement in the kingdom of Castile (1085-1350)', p. 54.

unique feature of frontier societies but was granted, along with other privileges like tax exemptions, only on the condition of military service. Despite all these measures, populating the frontier remained a troublesome affair.

From the 12th century on, regional leaders (kings) started to extend their power by consolidating areas into a contiguous territory and subjecting other authorities within that territory to rules and supervision.[32] One of the reasons behind this transformation was that the feudal system no longer provided the stability and security of earlier times. It had become a system that focused on property and land rather than on high values like fealty to a lord. The rising power of the kings cannot simply be explained by the comparative advantage of an extensive territory and its sizeable human resources that would make a ruler more successful in war.[33] For a long time, large kingdoms and small 'city-states' continued to exist side by side as equally resourceful political units. The background of the transformation to a central authority vested in the king was social legitimacy or the 'social empowerment' derived from the change.[34] The new form of power became associated with the provision of public order within its territory, and externally with statecraft aiming at the state's prosperity rather than at blind expansion. This ensured support from different strata in society. However, it was a very slow transformation that only in the 16th century made its impact felt on the European political scene, and even then it was still evolving. It is revealing that the jurist Jean Bodin (1529-1596) who introduced the term *sovereignty* in this period as 'the absolute and permanent authority of a commonwealth (state) who is not accountable to anyone else'[35], was amazed that the principle had not been formulated earlier (including by himself). What did Bodin have in mind? Many things of which we are now hardly aware because they are so obvious. As Paul Hirst succinctly lists them:

> Bodin's various marks of sovereignty – to give orders but not to receive them, to make laws, to administer justice, to coin money, to tax, to raise armies, to deal with other rulers – were complexly distributed across territory before the sixteenth century. Various agencies could do these things – including raise armed forces and enter into relations with other

---

[32] Piet Leupen, 1998, *Keizer in zijn eigen Rijk. De geboorte van de nationale staat.*
[33] As emphasized by Charles Tilly, 1990, *Coercion and Capital in European States, 900-1990.*
[34] John Gerard Ruggie, 'Territoriality and beyond: problematizing modernity in international relations'.
[35] *Les six livres de la République,* 1576.

rulers. The Hanseatic League, the monastic military orders like the Teutonic Knights or the Hospitallers, mercenary forces like the Catalan Company, city-states, bishoprics – all acted much as later 'sovereign' states would claim exclusively to do and often across the same territory.[36]

The Ottoman Empire
The Ottoman Empire (1326[37]-1922) displayed many of the characteristics that were mentioned in the previous pages: the recruiting of foreign troops, a decentralized system of rule and tax collection and at the frontiers either an increased degree of freedom for rulers (the marcher lords) who had to repel invasions from the outside or the creation of vassal states (tributary principalities). Like the Habsburg decision to place the Vojna Krajina in Croatia under direct supervision of Vienna, Suleyman I annexed a part of Hungary as a directly ruled province after 1540. The adjacent territories of Transsylvania, Wallachia, Moldavia and the Crimean Khanate remained under the rule of native dynasties that had to pay tribute to the sultan.[38]

All historic Empires struggled with distance and searched for ways to delegate authority to provincial governors or semi-independent principalities without losing control. When the Ottoman Empire was in its early stage and limited in territorial extension, the rule of new territories could be distributed among the sultan's family members. But simply being family does not guarantee solidarity, protection against a coup or even murder among its members. Both the distribution of power among family members and delegation of authority to pre-conquest dynasties suffer from the same defect: the enlarging of vested interests that run against the empire. Therefore, imperial governance always has to resort to counterbalancing strategies by mixing foreigners loyal to the sultan with lords that have hereditary rights in their areas. In the Ottoman Empire mixing meant that vested interests on one level had

---

[36] Paul Hirst, 2001, 'Politics: territorial or non-territorial?', The Global Site, First Press (website www.theglobalsite.ac.uk discontinued but texts still available 10/2/2018 on internet archive.
https://web.archive.org/web/20010620102350/http://www.theglobalsite.ac.uk:80/press/104hirst.htm
[37] Various years may mark the beginning of an Empire. Here we take the year of the capturing of Bursa by Orkhan which became the first Ottoman capital city.
[38] Colin Imber, 2002, *The Ottoman Empire 1300-1650. The Structure of Power*, p. 181.

to face centrally appointed functionaries (from elsewhere in the Empire) at a higher or lower level: provinces, districts (*sanjaks*) or fiefs. One important basis of authority on the lowest level was the awarding of fiefs to persons for services rendered to the Empire. Fief holders could live from the taxes on their lands and had to provide the sultan with military services.

The use of 'foreigners' assumed a very peculiar shape in the institution of slaves as members of the Sultan's household and military infantry (the Janissaries). Apart from conquering territory, Ottoman wars intentionally produced slaves among the non-Muslim (predominantly Christian) population. During the reign of Sultan Bayezid I (1389-1402) even a special body of raiders conducted razzias into enemy territory to satisfy the need for slaves. The members of this unit were recompensed with land and tax exemptions.[39] Contrary to the Western conception of slavery, these 'slaves of the Porte'[40] could reach very high positions such as minister or provincial governor. They were also highly esteemed as members of the military and perhaps less tormented by the 'home thoughts' mentioned in Li Bai's poem. It is an example of the fact that ethnic origin was less important in Empires than being non-corrupt, something guaranteed by a war captive's lack of local ties.

Compared to the Roman Empire and its European successors the administrative and military organization of the Ottoman Empire was highly efficient, as shown by its military advance in Europe. Occasionally religion was an important factor in maintaining the unity of the Empire, in motivating its drive towards expansion and in following a set of rules that to a certain extent protected the population of conquered territories (particularly those that voluntarily put themselves under the protection of the Empire). However, the rise and strength of the Ottoman Empire cannot be fundamentally explained by religious inspiration in the vein of *jihad* as some introductions to its history are inclined to claim. A typical explanation for the imperial drive of the early Ottoman state, in a book on Ottoman warfare, refers to the rise of militant Islam: 'This was the aggressive fanaticism of independent bands known as ghazis, groups of 'holy warriors' who fought to spread the faith and supported themselves through plunder (…) The early Ottomans were typical *ghazis*'.[41] This interpretation is at odds with the multicultural nature of the

---

[39] Ibid., p 132.
[40] 'the (Sublime) Porte' was a term for the Ottoman government used by French diplomats. It refers to the gate through which one reached the government buildings in Constantinople.
[41] Stephen Turnbull, 2003, *The Ottoman Empire 1326-169*, p. 10.

early Ottoman Empire, the many alliances between (Byzantine) Christians and the early Ottomans against common threats (of both Muslim and Christian origin) and the exchange relations (including intermarriage) across political boundaries between Byzantines and Seljuks (Osman's ethnic group). As Linda Darling suggests, the *ghazi* interpretations are based on the work of Ottoman historians from the 15th century, a period when new political facts promoted the adoption of a *ghazi*-identity.[42] Such facts were (among other factors mentioned by Darling): the Christian crusades and the advance of Ottoman conquests in the Balkan, where they met Christians who had no experience of cultural exchange with the Ottomans. These Christians intensified the Ottomans' awareness of religious difference and made them aware of the increasing number of commanders in the Turkish forces in Europe who were not of Ottoman origin and who might be motivated by stressing *ghaza* rather than dynastic aggrandizement. The emphasis on religious identity was also useful in cases where the Empire conquered neighbouring Muslim territories. Here it declared such aggression as a holy vocation to unite forces in the struggle against the infidel, a familiar drive in contemporary political affairs.

No opportunistic stance is foreign to an empire. For those familiar with European nationalism in the 19th century this 'invention of tradition' (as for example the *ghaza* tradition) sounds quite plausible, and in the Ottoman case it certainly shows the political utility of religion. One should be wary of explaining the Ottoman Empire's origins and impetus solely from an inherent conception of Islam.

The Colonial Empires

Separated from the 'Metropolis' by wide oceans and often started as enterprise by tiny maritime forces or commercial companies, colonial empires seem to represent a class of their own. They certainly did share some features with the classic empires, such as indirect rule by local nobility, the use of armies consisting of 'foreign' soldiers and a common religious identity engrafted upon European churches (although it never wiped out indigenous religious practice). Religious conversion was particularly apparent in the South American colonies where the practice of intermarriage between the

---

[42] Linda T. Darling, 2011, 'Reformulating the *Gazi* narrative: When was the Ottoman state a *Gazi* state?'.

European colonizers and the local nobility was another distinctive feature. Transport costs and time restricted the direct interference which land-based empires could afford. Only an extensive European settlement in the colony, including military reinforcements, could overcome such impediments over the course of time. That was only feasible in the 19th century. Initially, European overseas involvement focused on the gaining of light-weight but precious materials like silver and spices. The penetration of foreign continents was superficial and limited to coastal areas. Another factor that limited European influence was the encounter with old civilizations (particularly in Asia) that in certain respects were militarily superior to and more productive than the homeland of the visitors. The Industrial Revolution of the 19th century changed all that. It made the colonies part of the European economy, with a noticeable impact on their landscapes, plantations, mining sites, extraction of oil and wood products: even in places distant from the coast.

At this stage the territorial structure of a colonial empire like the British Empire was like that of ancient empires, with one major difference: the colonial empire's authority was established during the heyday of the national state. The metropolitan or core areas of these empires were densely packed in a European space where inter-state competition and mutual observation ran rampant. In a way, the potential destructive energy of this state system was deflected toward colonial expansion or support. It meant that boundary conflicts emerged in Africa and Asia rather than in Europe. They did not manifest themselves in the same way as conflicts did at the classic imperial frontier (a militarized frontier with continuously erupting wars). War was avoided either by creating buffer zones (such as Afghanistan) or engaging in speed diplomacy (i.e. the Fashoda incident[43] between France and Britain 1898). The unravelling of the colonial Empires did not start at the boundaries, but rather in the centre when European countries (with a major role for Germany as latecomer at the colonial scene) attacked each other in two 'World' wars. It enabled the imperial designs of Japan and unleashed (after the Second Word War) the decline of the (moral) hegemony of the imperial rulers. The loss of the empires was painful – not merely economically but also culturally – but the shock was softened by an unprecedented economic growth after the war. It sublimated imperialism into a belief in globalization that was just as gainful and perhaps less of a burden.

---

[43] The encounter between a French and British exploration team in search of colonial expansion in Africa.

Imperial territoriality: passive identity

Many of the features of empires mentioned above continued their presence after the Westphalian treaty when powerful states reconciled the idea of an international state system with colonial or postcolonial (US) expansion. Adam Watson even suggested that the history of world orders is rather a pendulum that swings between *empire* on one extreme and a world of *independent states* as the other extreme.[44] In between we find such categories as *dominion, suzerainty,* and *hegemony*. Whereas medieval Europe was primarily a complex of personal relationships that sometimes moved more or less as a whole when it was inspired by a religious idea (i.e. the Crusades), traditional Han China looked more like a regular state. While it lacked a public domain in which patricians could raise their voice about political affairs, which already existed in the contemporaneous Roman Republic, it had an intricate system of control that penetrated even down to the individual level, which Rome could not match. The base was constituted by small groups of five to ten persons in which everyone could be held responsible for the actions of the other group members. It necessitated the obligation to report any breach of norms by a member of the group to the authorities, which obviously could set family members against one another, something repeating itself in Mao's communist state.[45] Next to this existed an organization in households that mainly served to extract taxes and recruit labour for the great works of the Empire. Like in other empires, loyalty was created through a system in which orders of honour were conferred based on merits, which gave the recipient special privileges. However, this whole system did not create citizens since it was only organized **to dominate and not to listen to people** (apart from the juridical settling of disputes). Moreover, the inevitable transfer of authority to regional governors or kings always entailed the risk of disloyalty and attempts to seize imperial power. *Governance* is one of the great challenges in an imperial system, just like *closure*. Both are narrowly connected because an empire does not only struggle continuously with external but also with internal enemies. The boundary between these categories is ambiguous because external peoples are brought under the emperor's control and internal groups may convert into enemies. This forces the army to be equally focused on external and internal security,

---

[44] Adam Watson, 1992, *The evolution of international society*.
[45] Michael Loewe, 2006, *The Government of the Qin and Han Empires 221 BCE - 220CE*, pp. 135-150.

which increases the risk of disloyal commanders. The Qin and Han Empires took an effective security measure. They avoided making the institution of army commander a permanent appointment. The military rank of 'general' was allotted to reliable and courageous officials for the duration of a project, like the suppression of a rebellion or the pacification of nomadic neighbours.[46]

Despite such ambiguities, empires pursue clearly defined boundaries in order to enable administrative control and defence. Still, its subjects and officials represent their own emperor as a universal and divine ruler. There is no basic 'international' recognition of other peoples as sovereign tribes or states, only opportunistic attempts to deal with them when subjugating them turns out to be impossible. This conception of the empire as a natural world ruler is somewhat at odds with the revealed territorial practice of toilsome control (punitive expeditions) and closure (distinguishing areas with different types of relations with the Empire).[47] In order to solve this inconsistency, an imperial conception of the world should downplay the political relevance of ethnic identities. If everything is subordinated to the empire or the emperor, there can be no legitimate territorial right of others based on cultural distinction, at least not one that creates symmetric diplomatic relations. Neither is the territorial space that the empire controls subjected to cultural equalization. As in the case of the Roman Empire, the Empire is held together by an endless stream of victories in war that inspire awe among the conquered peoples. One may at best conclude that the classic Roman Empire did not merely rely on the stick but also on the carrot because it established cities or ports in conquered areas with economic gains as result. Further, this 'passive' status of identity literally made the utmost symbol of control in China, the Emperor, 'passive' by assigning him a role in a cosmic order that also penetrated daily life in the village. The Emperor had to perform rites, just as the villagers did, in order to preserve the local harmony. The awe of the Emperor was inseparable from awe of the cosmic order, a vital energy that permeates everything

---

[46] Ibid., pp. 58-59.
[47] Thus, the dynamics that characterize 'imperial territoriality' is located in the *governance* category because Empires suffer from limitations in bridging distance and in the closure category because a stable framework for dealing with other powers is lacking. An almost caricatural example of the latter are the messages between imperial rulers like Pope Innocent IV and the Mongol ruler Möngke Khan (a descendant of Genghis Khan), handed down from the middle of the 13th century. These messages contained high-flown demands to submit to the sender without allusions to a future compromise (Voegelin 1940-41, p. 39).

and, although it creates all kinds of spatial distinctions (like specified in the *Feng Shui* system), it is **de-territorializing the Emperor as a territorial sovereign** (Son of Heaven).

Table 2.1 Imperial territoriality

Dimension	Active/passive	(Un)boundedness
Closure Structure of authority. Who are (un)affected by dominant authority?	shifting internal and external boundaries (active)	area of pacification centre-periphery frontiers
Governance How is a territory / group engaged in turning it into a resource?	struggle for power (active)	personal power decentralised to lords or governors coercion (army / war) laws against abuse of local authority
Identity How do people attribute *meaning* to power in (territorial) assemblages?	unchangeable cosmic order (passive)	unity under one divine emperor

The European medieval central authority was unique in being split between an emperor and a pope, but both were seen as divinely sanctioned. The Ottoman sultans failed in acquiring a sacred status although their centuries-long rule as a single dynasty intimated the idea that God favoured them. It was, however, the very reverence for long genealogical links in the Turkish and Mongol tradition that proved the Sultans' ineptness to claim descent from the Prophet or to assume the status of Caliph, which was reserved for the Prophet's Quraysh tribe.[48] De-facto power over the Islamic world, and particularly being the guardian of the holy cities Mecca and Medina, made the Sultan's status equal to the earlier Caliphs, although this title was not always officially used. The legitimacy of an empire does not necessarily depend on an outright divine status of the emperor. The sheer success of the imperial enterprise or the adoption of the status of defender of the faith (the later *ghazi*-identity of the Ottoman Sultans) may also have sent the message that God agrees with its authority. When from the late 18th century on the Ottoman Empire crumbled away and its Muslim subjects succumbed to foreign rule in the Crimea, the Balkans, the Caucasus and North-Africa, these people

---

[48] Hakan T. Karateke, 'Legitimizing the Ottoman sultanate: a framework for historical analysis'.

and even those as far away as the Muslims of South-Asia experienced a new urge to recognize the Sultan as their spiritual leader because he represented the only independent Muslim power in the world that could stand up for their interests. As a kind of exchange for being recognized as extraterritorial protector of the Muslims in the Russian sphere, the Sultan agreed with Russian protection of the Christians in the Ottoman Empire. This emphasis on identity and diplomatic exchange also indicates the transformation to **the new territorial order of centralized states**.

The post-Roman European realm showed the strongest degree of deterritorialization. The dualistic figure of leadership (emperor/pope) ensured the people's gaze up to heaven, and actually filled the earthly world with miracles and manifestations of the Holy Spirit. The ecological continuity of its margins and the more personalized and diffuse system of authority in medieval Europe made the frontier easier to manage than in China. Boundaries did not play an important role and Europe's spiritual centre, Jerusalem, was even located beyond the imperial frontier! The split between religious and worldly power made the idea of authority as something exercised from a geographical centre less obvious still. It resulted in a paradoxical outcome: kings took on responsibilities for governing tasks that had been neglected and they subsequently developed them into **central powers** around which all other things moved ("sun king" as a later French king was aptly named). When the first signs of this development took shape, religious scholars began to declare more emphatically that the position of Jerusalem was 'in the middle' of the world'![49] It was a desperate incantation against the unstoppable territorial fragmentation.

Notwithstanding the great differences along the continuum running from the weak Carolingian territorial structure of Europe to the state-like Ming Empire in China, there is common ground in the way empires deal with territoriality. First, they are never sure how to deal with the transition between an inner and outer sphere, precisely because they consider themselves as the only legitimate source of authority on earth. 'On Earth, there is only one Lord, Chinggis Khan', ran the preamble to the letters that the Mongol rulers sent to pope Innocentius IX who had complained about their cruelty: 'You, who are the great pope, together with all the princes, come in person to serve

---

[49] Iain Macleod Higgins, 1998, 'Defining the Earth's center in a medieval multi-text. Jerusalem in The Book of John Mandeville'.

us'.⁵⁰ The discrepancy between the ideology of universal rule and the reality of limited power makes *closure* (who are the emperor's subjects?) a source of continuous worry in empires. Second, this is also valid for internal control (*governance*), which suffers from the sheer extension of the imperial surface and inevitably depends on the delegation of authority to local governors. All emperors have devised information systems to monitor the quality of 'local government', but it never could match the control of a centralized state. Imperial territoriality pivots on *governance* and *closure* whereas *territorial identity* only applies to the ruling elite. The medieval European Empire may have benefitted from an unusual cultural unity, but by linking this identity to a separate religious leader, the pope, the worldly Empire was in the end territorially fragmented rather than consolidated.

This is the way the world ends...

'**N**ot with a bang but a whisper' predicts T.S. Eliot in his poem 'The Hollow Men', and it seems at least valid for empires. They did not collapse but languished, most likely without the shocks that unleash a nervous search for new meanings. Ordinary people were not suddenly exposed to economic or physical insecurity. The dominance of daily worries about crop failure, epidemics and the arbitrary deeds of local lords will have taken place first. Those most likely to feel the impact of a changing territorial order were themselves linked to the system of governance, either in a worldly or spiritual sense. A thing to keep in mind is that one also needs to experience the twitch of a new territorial logic to be put off one's balance. This was the case for intellectuals like Dante Alighieri who clearly regretted the whining of the imperial order

In 1302 Dante was found guilty of corruption during his term as city-prior (the highest function in city government) in Florence and condemned to perpetual exile. The accusations started after one of those periods of infighting that characterized the political culture of the Italian city-state (see also Chapter 3) and in which the Black Guelphs had eliminated the White Guelphs. Dante's indignation about his fate and the politically motivated false accusations against him led him to write a poetic work, the *Commedia*, that was both a denunciation of the political establishment of Florence and a reflection on the right juridical order, thereby shifting the emphasis from city

---

⁵⁰ Sh. Bira, 2004, 'Mongolian Tenggerism and modern globalism'.

to the empire as source of legitimate government.[51] Dante took delight in sending one after another Florentine dignitary to hell in his story.

As precursor of the future state, the Italian city did not bode well for the mechanism of justice in a world of sovereign states. In *De Monarchia,* Dante puts forward the necessity of a supreme authority like that of the classic Roman Emperors who proved their legitimacy, in Dante's opinion, by their capacity to exercise superior authority on a large scale. For Dante, who did not endorse the role of the pope in worldly affairs, this capacity to rule automatically implied divine legitimacy.[52]

The mental difficulty produced by the transition from an imperial mode of territoriality to the world of states resulted from the new activation of the identity dimension: the claim that territorial units are legitimate because they join people with a common interest together. Not that identity or territorial attachment was absent in late medieval Italy. Dante was full of his Florentine identity to such an extent that his exile almost felt like sudden death. However, territorial identity as a political construct, as an expression of legitimate common rights and political aims, was difficult to separate from the struggle for power of the day, which evoked identities of gangs or families rather than a wider interest. The question 'to whom does the state/city belong?' pushed aside all urges to produce a just government. Ultimately, Dante's solution of reverting to the past, to the beneficial reign of a superior power, was not viable. New social realities require new ways of adapting both in a technical and a mental way.

---

[51] Justin Steinberg, 2014, *Dante and the Limits of the Law.*
[52] Stuart Elden, 2013, *The Birth of Territory,* pp. 189-193.

Chapter 3
A NEW JERUSALEM: THE BIRTH OF THE TERRITORIAL STATE

> O David, thou soughtest shelter
> From King Saul's tyranny.
> Even so I fled this welter
> And many a lord with me.
> But God the Lord did save him
> From exile and its hell
> And, in His mercy, gave him
> A realm in Israel.
> **Wilhelmus** (Dutch national anthem), ca 1570, 8th verse[1]

> All the realm [of France] being one great city,
> after the fashion of the Roman Empire.
> **Jean Bacquet** (1580)[2]

Holy ground and ordinary people

Early modernity brought the dissolution of empire in Europe and replaced personal authority with territorial authority. The simplicity of this statement downplays the difficulty experienced by contemporaries to understand and accept the new reality. Post-imperial polities or states were not pre-existing entities shaking off their imperial chains. Such entities or nations were characterized either as dynasties with unstable territorial ties or as new entities that had to be created 'from scratch'. This re-territorialization required a cumbersome mental switch: accepting territorial identity as a legalizing factor. Today it is difficult to understand why the leader of a war of 'liberation' would

---

[1] I could not retrieve the name of the translator, but this widely distributed English version of the *Wilhelmus* invariably contains an error in precisely this verse where it states: 'But God the Lord did save *me*' instead of '...*him*' (as corrected in this quote). This error is somewhat understandable because the poet intended to conflate the history of Israël and the Dutch Republic.
[2] Charlotte C. Wells, 1995, *Law and Citizenship in Early Modern France.*, p. 31.

worry about the legitimacy of his acts, even if it expresses the wish of a people. Territorial 'self-determination' was – even stripped of all the 20th century democratic requirements – an alien concept in the early 16th century. One of the seemingly unambiguous liberation struggles of 'a people' in the 16th century, the Dutch Republic contra the house of Habsburg as remnant of imperial Europe, reveals the struggle to imagine 'self-government'. The Dutch national anthem still contains an embarrassing lamentation of the first Dutch leader, Prince William of Orange (1533-1584), about his loyalty to the Spanish (Habsburg) king in which he seems to regret the imminent rift. Moreover, those who revolted against distant or indifferent rulers at the dawn of the Modern Age were not necessarily attacking the principle of personal sovereign authority. As the histories of early revolutionary nations like the English and the Dutch show, after having done away with a bad king the political reflex was to seek shelter under the wings of another (foreign) sovereign. Self-government could not be imagined in a different way.

The essential mental switch that was required involved the conversion of identity from a territorially *passive* to a territorially *active* principle. How difficult it was to make such a step at the end of the imperial Middle Ages is illuminated by the fact that religious discourse – the very embodiment of de-territorialized identity – was cunningly deployed to conjure territorial identity. The reference to the biblical King David in the Dutch anthem suggests that God may give a territory. It illustrates one of the mechanisms mentioned in Chapter 1: producing 'a new (emergent) reality on the basis of pre-existing agencies and resources'.

There are indications that a rudimentary idea of national feeling was already cropping up in the 12th century, although it could not yet be articulated by any intellectual means as in the later Modern age. The medieval historian William of Malmesbury, writing the history of English kings around 1127, encounters an awkward event when he has to deal with the invasion of William the Conqueror (1066).[3] William's victory created a new political space, Normandy and England, and a new ruling class in England. Malmesbury acknowledges that 'blood of either people [Norman and English] flows in my veins', but he still looks for ways to meaningfully particularize the English territory and its people. The only intellectual means available for such an en-

---

[3] Robert M. Stein, 1998, 'Making history English. Cultural identity and historical explanation in William of Malmesbury and Lazamon's *Brut*'. In: Sylvia Tomash and Sealy Gilles eds., 1998.

terprise was Christian religious discourse. Malmesbury indulges in a tour of English places that had witnessed miracles before the Norman Conquest – stories about saints whose bodies were found 'entire after death typifying the final state of incorruption'. The 'body that does not decompose' is an obvious symbol of the country and Malmesbury's tour around different sites of English miracles adds in its own way to the idea of a sanctified territory.

Some of the stories he tells are also rich in other symbolic allusions. A young boy, successor to the throne of the Mercians (one of the English tribes), is murdered in a distant forest at the instigation of his sister who desires the throne. The crime that has remained hidden is nevertheless revealed because a pigeon drops a parchment scroll in which the story is told on the altar of St. Peter in Rome. Nobody can read the letter because it is written in English, but an Englishman who happens to be in Rome reveals the message. He translates the text so as to enable the pope to send a letter to the English kings in order to inform them about the crime. The story underscores the location of England at the periphery of the Christian world ('situated almost out of the world' as Malmesbury says himself) but shows at the same time that it is possible to do good works by translating Latin, one of the holy languages, into the English vernacular. Christian geography only allows a privileged position to Jerusalem, but the fact that the Lord has seen fit to perform such miracles in England demonstrates that no one should despair about the reality of the Second Coming of Christ, not even in a peripheral place. In this roundabout way Malmesbury declares the English territory to be a place for human fulfilment involving God, kings and people. Saints and translators (like the writer himself) are the media that knit this world together in a meaningful way.

Malmesbury was a secluded intellectual engaged in constructing a national identity with at most a *latent* political significance. There is no reason to believe that he told something that strongly occupied the mind of the masses. For an early voice of ordinary people, we should rather look at the dynamic city-states of Italy, where a clever individual with a sense of enterprise could have an impact, whatever his background.

Lauro Martines portrays the Italian city-state of around 1300 as a territory that arouses 'intense love and hatred'[4]; the kind of feeling that one promptly associates with nationalism although the term would be inappro-

---

[4] Lauro Martines, 1980, *Power and Imagination. City-States in Renaissance Italy*. pp. 111-112. Subsequent quotes are taken from these pages.

priate in this historical period and setting. A man exiled from Lucca vowed that if he could ever return he would '…go licking the walls all round and every man I meet, weeping with joy'. The poet Guittone d'Arezzo, a voluntary exile who had left his place of birth Arezzo after political antagonists had seized power, writes in a poem that '…an evil base regime /…/ make me hate my land/ alas, and love another's'. There was good reason for the tug of local feeling, writes Martines. People always had grouped together in places to find protection and help but 'as the commune expanded the life of urban residents came to turn more and more around the fortified buildings of local government. In this atmosphere the New Jerusalem seemed to some men to lie only inches away, if they could but take control of government'. What the poet Guittone d'Arezzo reveals is a new sentiment that links the quality of a place ('hate my land') to the quality of the government ('an evil base regime'), which usually meant that the government should represent one's narrow interest group. Anyhow, this politicization of the community created the first public space, a field of communication and political comment, in Medieval Europe. In the Italian city, internal divisions sensitized political nerves, but it never engrossed the entire community

As the urban landscape of the old cities in Northern Italy and Tuscany unmistakably show, the early political history of these communities was not merely a story of united patriotism. The tall and inaccessible towers were strongholds of noble families or clusters of families who at the end of the 12th century were in a permanent state of war. After a period of consensus in the 11th century (the 'consulate'), in which all political energy focused on the liberation of the city from external influence (other cities, bishops, the pope, or the German emperor), internal struggles again took over the political scene. The neighbourhood (or still smaller the *vicinanza*) was not only the natural scene and centre of most social life but also a forum for politics, government and public life.[5] Two developments bound these neighbourhoods more firmly to the city as a whole: the expansion of city-state authority and the emergence of the *popolo*.

Early political life excluded a large group of town residents. Partly they were people who could not boast of impressive names or family traditions but who had gained self-respect and wealth from trade or craft. Another group, attracted from the outside by the promise of urban progress and a demand for personal services, found themselves again in a position of de-

---

[5] Philip Jones, 1997, *The Italian City-State. From Commune to Signoria.*, p. 404.

pendency where 'dynastic' loyalty to a noble or rich family counted as the utmost virtue. In this respect urban life had not completely replaced the vertical relationships of a feudal order with a territorial perspective. This changed from the moment that, around 1200, the *popolo* burst upon the scene: a group of assertive townspeople that did not feel politically represented. They were recruited from the middle-class, not from the poor or the servants. With the guilds, which were craft associations that sometimes were even armed, as vanguard they succeeded in claiming a third to a half of the seats in city government. In Bologna, Milan and Florence the city was subdivided in districts each controlled by an armed company of the *popolo*. Sometimes they geographically overlapped, but since each member and house was uniquely assigned to a company no confusion arose about responsibilities. With a similar administrative thoroughness, a governmental organization was erected with eight to twelve 'aldermen' at the top and a legalistic system of contracts, notaries and judges. These agents forged the social and material world together in such a way that trust and predictability, essential conditions for a market and growth economy, were established. The time-honoured nobility looked on it as the work of a shadow government, a fictive community, but this was a gross denial of the actual power wielded by this system in the urban goings-on. The nobility failed to realize that it revealed a glimpse of the modern politico-territorial order.

The most important effect of the shifting political scene on the individual was a mental appropriation of the entire urban space and a new self-awareness based on the idea of being a stakeholder in a collective enterprise. Bocaccio describes how a rich Florentine baker hesitates to address a nobleman who frequently passes his shop, but when at last he overcomes his fear and recommends the merchandise, he does not assume a subservient attitude.[6] Much later in history a similar scene would be called up in the setting of a modern autocratic state. In Chernyshevsky's novel *What is to be done?* (1863), the leading character, a representative of Russia's 'new people', encounters a noble pedestrian in the streets of St. Petersburg.[7] As they follow the same track one of them is obliged to get out of the way. But it is not the man without recognized status that clears the way for the established and dignified gentlemen. Ultimately his noble adversary lands up in the gutter. Of course, members of the higher classes could avoid such rude confrontations

---

[6] Lauro Martines, 1980, p. 63.
[7] Marshall Berman, 1982, *All that is solid melts into air. The experience of modernity.*

by making use of coaches and sending servants on errands, but that didn't alter the fact that urban space had been democratized in a way. This was also manifested in the layout of the Italian city where public life became refocused on the public palace(s) and main squares. Even churches became part of political life by accommodating the rites associated with the installation of new 'mayors', for example by making them burial sites like the Venetian doges did with the San Marco or by keeping the relics of the local patron saints.

The fusion of politics and religion is one of the curious characteristics of the Italian city-state that can be seen as directly linked with the ambiguities in its political life and structure. Philip Jones, pointing to the emphasis on theology and the ideals of monarchy and empire in medieval political thinking, aptly summarizes the problem: 'Republicanism [of the Italian city-state] in doctrine as in practice had to make its way in an alien environment'.[8] On the one hand this meant that republicanism remained an ideal insofar that the communities evolved as untidy federations rather than as political unities.

> From its origin and by its very nature, collective and participatory, the commune evolved not as a unity but as a federation, a loose and widening association of semi-autonomous, private forces, old and new groups, 'societies', and alternative comunia. And among these, despite all unitary state-building, citizens only too readily divided their allegiance.[9]

On the other hand, it meant that attempts to regulate the political system reached for law and moral indoctrination rather than for models of representative government. Critical observers of the day did not latch onto failures like political corruption in order to start a discussion about the political system – they merely associated such deficiencies with *individual* vice: avarice, ambition, usury.[10] Vices should be dealt with by the church and religion. The church (or rather Christ) as provider of redemption from sin seemed to be the appropriate power to consecrate the political body of the city that was even compared with the 'mystic body of Christ'. The term New Jerusalem was not merely a metaphor; it was taken up in redesigning certain parts of the city like the St. Mark's Square in Venice and, 'in accordance with the heavenly city', in giving cities a circular or cruciform shape.[11]

---

[8] Philip Jones, 1997, *The Italian city-state*, p. 460.
[9] Ibid., p. 540.
[10] Ibid., p. 539.
[11] Ibid., p. 298.

In Italian cities the aim of separate groups was not to occupy a proper and dignified place in the state but to capture the state. As a response to increasing instability at the end of the 13th century, despotic tendencies returned, and were even welcomed, in many an Italian city. The Venetian doges are the best-known example of heads of city-government that adopted almost kingly authority. Aware of legitimacy problems in a fragmented community, they took care to involve the religious sphere in order to stress the bonds with and within the community. Andrea Dandolo, elected Doge in 1343, was highly successful in this 'blend of the bureaucratic and transcendent'.[12] Architectural changes and artistic embellishments of San Marco Cathedral, particularly focusing on the baptistery, constituted the heart of his efforts. Dandolo was administrator of San Marco (before his appointment as doge) when the ground-breaking decision was made to erect the tomb of doge Soranzo (1312-1328) in its baptistery. From the 12th century onward, baptistery building was part of the celebration of communal identity. It was the site of entrance into the local Christian community: a symbol of both civic and Christian identity.[13] Burying the rulers in the baptistery associated them with the baptism of Christ, a ritual in which man is infused with the divine spirit.

There was another more direct way to link the community with the divine sphere – through the cult of the patron saints.[14] In each cathedral, relics of one or more saints were preserved. Usually these were the remains of a bishop from the early days of Christianity, preferably sacred personages who were native and exclusive to the locality. Such saints represented both a communal power that could be unleashed by processions and worship, and the historic continuity of the community. By walking in processions in which the relics of a saint were paraded, the community symbolically governed the city. On a few occasions the explicit aim of the procession was to avert a real danger such as drought, an epidemic or an approaching enemy. Whatever the motive, participation was compulsory and the arrangement of the procession in groups from each neighbourhood underscored the link with the secular functions of the city. Fines for non-performance and incentives for informers about cases of negligence were arranged as well. Some members of the clergy

---

[12] Debra Pincus, 2000, 'Hard times and ducal radiance. Andrea Dandolo and the construction of the ruler in fourteenth-century Venice'. In: John Martin and Dennis Romano (eds.), *Venice reconsidered. The history and civilization of an Italian city-state, 1297-1797,* p.94.
[13] Ibid., p. 99.
[14] Diana Webb, 1996, *Patrons and defenders. The saints in the Italian city-states.*

preferred to put the practice in a more mythical light. At the end of the 12th century Bishop Sicard of Cremona remarked about the Palm Sunday procession: 'Perhaps we are recalling in this procession the procession of the children of Israel, who on this day crossed the Jordan with dry feet…'.[15] He referred to the Bible passage in which the wandering Jews finally reach the Promised Land aided by God who cuts off the waters of the Jordan (Book of *Joshua*). Holy ground was certainly a justified label for the way Italian people looked at their cities during the 12th century and later. But explicit comparison with the events of the Old Testament was something new and almost superfluous in the presence of such divine, but at the same time manipulatable, local saints. The theme of wandering or moving was well chosen in a society that was territorially expanding in the *contado* (countryside), engaged in exploring new trade routes far away and struggling with a faltering (Holy Roman) empire over rights and competencies. Such conditions repeated themselves again, but more shockingly, across the Alps and in North-Western Europe in the 16th century. *Identity* was on the move.

Making identity territorially active

Each medieval attempt to legitimize territorially discrete government needed to 'write' politics as part of an earthly playfield of divine forces. Territorial bundling of authority, however, made this an increasingly ambiguous activity because it seemed to take away the authority of those central representatives of the Divine Empire: the Pope and the Emperor. In the early 12th century Louis II for the first time used the title *rex Franciae*, King of France instead of King of the Franks. This lexical shift from personal relations to the territorial[16] also implied the assurance by the King that he would not pursue his own interests but represent the commonwealth or *res publica*. At that time this involved supervision of rural landlords and town leaders rather than the mass of the population. The latter were woven into the territorial identity by emphasizing the holy character of France. The French king (Louis VI, 1214-1270) was 'the most Christian king' and his people 'the chosen people'. Outsiders like Pope Clement V (1304-1315) also applied these qualifications to France, but French kings applied them most fervently themselves. They ad-

---

[15] Ibid., p.19.
[16] Georges Duby, 1991, *France in the Middle Ages*.

vanced the idea that God especially favoured the kingdom of France and that it was the most important part of the Church. Any attack on the king would be an attack on the faith. It was a geopolitical image that served the Capetian kings in claiming an independent role versus a German Empire that posed as the successor of the Roman Empire and the pope[17]. The Germans were painted as the 'Evil empire', as a Dominican priest suggested at the time in a pun with *'en pire'* (literally: worse). Moreover, it casted the 'devout' French common people as an ally – the only way to involve the people in 'politics' that fitted the spirit of the times.

There was another way to mentally accept the territorial bundling of authority and the operations of worldly governments exercising moral authority in society: It simply involved the recognition of a basic divide between the divine and earthly spheres. If human beings could never substantiate the claim that events on this earth were a direct result of divine action, recognition of multiple governments would suddenly be easier. The only obligation of such governments was to see to good governance. This reflected the philosophical teachings of Martin Luther (1483-1546) and Desiderius Erasmus (1466-1536). Erasmus made some interesting remarks on the links between the spiritual sphere and worldly communities such as cities. He opined that 'towns ruled by priests stagnate and fall into ruins',[18] but he also had written, 'What else is the city than a great monastery?'[19] The latter was not a plea for theocracy, as the preceding quote makes unambiguously clear. It was an ode to people who in all equality are working for the common good (*res publica*) guided by a utopian vision about heaven or earth. It also suggested that the quality of government depended on the right attitude of people; governance was indeed a matter of keeping the moral order. Luther broke with the medieval epistemology of Thomas à Kempis, which was based on the assumption that for the true believer God's works are clearly recognisable in worldly events.

Kirstin Zapalac has provided a striking illustration of the epistemological turn that Luther's message induced in the 16th century.[20] She found a

---

[17] Joseph Strayer, 1971, 'France: the holy land, the chosen people and the most Christian king'. In: Joseph Strayer, *Medieval Statecraft and the Perspective of History*, pp. 300-314.
[18] James D. Tracy, 1978, *The politics of Erasmus.*, p. 29.
[19] John Merriman, 1990, *A History of Modern Europe. Volume I.*, p. 104.
[20] Kristin E.S. Zapalac, 1990, *In His Image and Likeness: Political Iconography and Religious Change in Regensburg, 1500-1600.*

Chapter 3. A NEW JERUSALEM

Figure 3.1. Good governance. Isaac Schwentner, "*Das gute Regiment*" 1592 [The good government]. © Museen der Stadt Regensburg, Peter Ferstl.

painting representing a city council meeting of the German town of Regensburg in 1536. At one wall of the room in which the meeting is being held hangs a painting that shows 'The Last Judgment' as an event happening on Earth. This image was recurrent in places where justice was administered from the later Middle-Ages onward. Another painting done almost one hundred years later, in 1627, shows a similar council meeting in this very room. On exactly the same place where 'The Last Judgment' had hung in the earlier picture, a new painting decorates the wall. It is partly covered by a curtain, but enough is visible to recognize it as a work of art that is still preserved in our time. It is a painting from 1592 by Isaac Schwentner in which allegorical figures exemplify the various virtues of good government (Figure 3.1). A small portion of the painting located in the upper sphere also shows the Last Judgment. This scene simultaneously shows a connection, yet also a clear separation, between the Last Judgment as a divine event and the operations of (good) government in this picture.

The belief in the 'transcendent' nature of divinity rather than in the divine presence in the world was not limited to Germany, the heartland of the Reformation, but also extended from the year 1530 into England as a very deliberate and conscious process. It was accompanied with material changes like the seizure of Church property, monasteries and land, which in some cases (such as Kent county) accounted for 40 percent of the surface area.[21] Such transformations completely changed the experience of space. Maps published after 1570 started to show a completely different landscape: 'the old landmarks – the paths of pilgrimage, the shrines that had tapped into grace and health, the monasteries that sheltered travellers and beggars – did not appear on these maps'.[22]

It is customary to see the Reformation as a booster of secessionism because Protestants resisted religious coercion by Roman Catholic rulers and ultimately claimed the principle '*cuius regio, eius religio*' (like sovereign, like religion) from the Empire. In the above, however, we may discover a different chronology: how the epistemological turn of the Reformation was inspired by the territorial bundling of authority (sovereignty) that was already in progress. The rise of the sovereign state shocked people because it seemed to drive a wedge between them and the road to eternal salvation. Only by ecstatic appeals to the holiness of the nation did French kings of the 12th century manage to allay this fear. But they could not deny that their state's intention was to swallow authority of local rulers, bishops and the imperial representatives of God. By arguing that no earthly agent could really claim to be a man-

---

[21] C. John Sommerville, 1992, *The Secularization of Early Modern England: From Religious Culture to Religious Faith.*
[22] Ibid. pp. 25-26.

ifestation of God's actions, Lutherans found another way to legitimize a type of government that was not sanctioned by the highest authority on Earth. The Reformation was a spiritual answer to the territorial shock of the emerging state rather than the reverse.

Of course, crossing out external legitimacy immediately caused another problem that was not easy to solve in the absence of an international system with universal rules of engagement (*closure*). In such a situation religion could still provide a compass, but only by orienting people toward events that according to the Bible had really happened on earth or were going to happen (the Apocalypse). We already met the favourite Exodus story about the wandering Jewish people in a Cremonese priest's sermon. The story would repeatedly turn up in the powerful early-modern Dutch and English states that both struggled to attain a moral conception of their place in the world and territorial reach. The unlinking of identity from a static vertical chain that reached into heaven automatically put the horizontal issue of closure into the limelight. Therefore, I would characterize the mental struggle of early-modern territoriality as work on a *closure-identity* compound. How we have to conceive the role of governance in the early-modern age will be discussed later in this chapter.

Exodus and geopolitics

Whereas the Italian city-states were an internal nuisance for dynastic rulers over Europe, the formation of the Swiss Confederation around 1500 and the Dutch Republic later in the century was a shock to the wider world as well. It showed the force of local alliances against apparently overwhelming powers and it foretold the fall of everything that dynastic rulers had taken for granted in governing territories as a family business. But their success and the need to constitute a unified political body was just as much of an embarrassment for the emerging states themselves, particularly the Dutch Republic with its global trade aspirations. Its official name was The Seven United Provinces, henceforth called the Dutch Republic. The term province is somewhat misleading as it suggests the kind of regional subdivision that constitutes the administrative framework of a centralized state. The provinces of the Dutch Republic should rather be compared with the units of a federal state. They were sovereign polities on their own terms but had transferred some competencies like foreign policy and defence to a federal body, the 'States General'.

Chapter 3. A NEW JERUSALEM

Some writers have tipped the scale to the other side by stating that the Republic was merely an ad hoc conglomerate of independent states with seven armies etc. This would never have resulted in its successful resistance against the militarily strong Spanish (Habsburg) state and the rise of a Dutch hegemonic world power in the 17th century[23]. Each province paid a fair part of the total defence costs but the distribution of soldiers and army units across the territory did not spatially correlate with the financial input.[24] More than half of the revenues for military purposes went to the 'Generality lands' (territory won over from the Spanish but directly governed by the States General) whereas the Union also occupied some cities outside its confines. In the official document that initiated the Union (Union of Utrecht, 1579) not much is specified on the philosophy behind this state structure and even many of the federal arrangements (covenants) were made in practice rather than set down in a written constitution. But one can safely establish that the very resistance of the people in this corner of Europe was to a great degree motivated by aversion to the centralising spirit in the kingdom or empire from which it broke away.

The personalization and glorification of Habsburg power can be understood from Sebastian Münster's often reproduced map of Europe (1588). In this map, Europe is represented as a queen, with the Habsburg possessions associated with her vital body parts. It would be illogical to expect a recurrence of this hierarchical vision on the domestic affairs of the newly independent and Protestant Low Countries even though some parts, like the province Holland, were more rich and powerful than others. Notwithstanding the fact that Holland was the core region of the new Republic, the province was itself a conglomerate of cities and regents, which occasionally split in political factions and could even ally with those with common interests in other provinces[25]. As an institution neither the States General nor Holland were comparable with the governments of other emerging states. At all administrative levels in the Republic it was normal to give authority to a council rather than a person. Even the most conspicuous person in this state, the *Stadholder*, was subjected to the system of checks and balances. He was the

---

[23] Peter Taylor, 1995, *The way the modern world works. World hegemony to world impasse*, p. 51.
[24] H.L. Zwitzer, 1991, *De militie van den Staat. Het Leger van de Republiek der Verenigde Nederlanden*. [The Militia of the State etc.].
[25] Simon Groenveld, 1990, *Evidente factiën in den Staet. Sociaal-politieke Verhoudingen in de 17^e-eeuwse Republiek der Verenigde Nederlanden*. [Revealed factions in the state etc.]

continuation of a similar authority that in Burgundian times had acted as the local representative of the distant sovereign.[26] Since William ('the Silent') of Orange had led the rise in arms, most provinces assigned this function exclusively to members of the House of Orange. The *Stadholder* of the Republic was commander-in-chief of the army and navy and had the right to elect urban magistrates and appoint provincial officials. As Kossmann notes, 'the presence of a princely tradition helped to mitigate the absolute character of the ruling oligarchy, though it should be added that the princes of Orange were rarely willing and never able to supplant the Holland plutocracy from whom they and the Republic ultimately received the money they needed and the directives underlying their policies'.[27]

The project of unifying all the Low Countries, meaning the contiguous area of seventeen provinces that had been ruled by the house of Habsburg and already knew a kind of local representation, the *States General*[28], never succeeded. After forty years of armed struggle, characterized by sieges and resistance of cities rather than pitched battles, both adversaries realized that they had reached the limits of their capacities. An extension of the frontier southward would create a vulnerable line of defence requiring an army size that could never be recruited by the Republic, in addition to the fact that the increased wealth of the Dutch had diminished the inclination for continuing military sacrifices. To the Spanish armies the aquatic defences of the core provinces of Holland and Zeeland appeared unconquerable, and in 1609 a cease-fire was concluded that would last for twelve years. In actuality, this cease-fire decided the territorial separation of the Dutch Republic and the 'Spanish Netherlands' and was ultimately sealed by the Peace of Münster (1648). But just like the boundary between Spain and France, this new boundary could not be accepted by the military as a line that would settle all international relations (see below). After 1667 the pattern of European alliances changed in a radical way when France emerged as a greater threat to the Dutch: the Spanish Netherlands took the form of an indispensable buffer area. The Brussels representative of the Spanish government agreed in 1698

---

[26] A. Th. Van Deursen, 1999, 'The Dutch Republic, 1588-1780' in J.C.H. Blom & E. Lamberts, *History of the Low Countries.*, p. 149.
[27] E.H. Kossmann, 2000, *Political thought in the Dutch Republic. Three studies.* Amsterdam : Koninklijke Nederlandse Akademie van Wetenschappen, p. 16.
[28] Established by the Burgundian duke Philip the Good in 1464, in view of better responding to foreign competition.

Chapter 3. A NEW JERUSALEM

to garrison eight Spanish fortresses along the French frontier with 8000-9000 Dutch troops.[29] This role of the Spanish Netherlands as a barrier (supported by the great powers Britain and Austria) continued unto the middle of the 18th century. All the time the Dutch representation of its national territory was ambiguous, as demonstrated by the popularity of the map of the seventeen provinces continuing until 1800.[30] Twenty versions of these wall-maps are still known against only four versions of the *seven* provinces. The latter, not surprisingly, started to appear after 1648.[31]

How then did the mental appropriation of federal territory proceed given the strong independence of the separate units? The impression of most historians is that any impetus from the side of the rulers to arouse nation-wide patriotism was missing in the Dutch Republic.[32] The dominant form of political identity was local or regional patriotism. However, in moments of crisis or revolt, the name of the House of Orange could arouse collective feeling. Rituals of the state like the opening of the States General or glorious entries of the *Stadholder* could have the same effect if they reached the population through media like pamphlets or public plays. The degree of literacy was sufficiently high[33] to expect a unifying force from the printing press, but

---

[29] Olaf van Nimwegen 'The quest for security: the case of the Dutch Republic'. In: Michael Burgess and Hans Vollaard, 2006, *State Territoriality and European Integration*, pp. 17-36.

[30] H.A.M. van der Heyden, 2001, *Kaart en Kunst van de Zeventien Provincies der Nederlanden* [Map and Art of the Seventeen Provinces of the Netherlands]. Similar conclusions have been drawn from an analysis of all maps in smaller printed form and atlases. See H.A.M. van der Heijden, 1998, *Oude kaarten der Nederlanden 1548-1794 / Old maps of the Netherlands 1548-1794.*, p. 98. See for an overview of the changing shape of the Netherlands: Hans Knippenberg, 1997, 'Dutch nation-building. A struggle against the water?', pp. 27-40.

[31] These facts suggest that Simon Schama is pushing his interpretation too far in suggesting that the large wall-map of the seventeen provinces that can be seen in Johannes Vermeer's painting 'The atelier' expresses Vermeer's Catholic nostalgia for the undivided territory of the Republic. Moreover Vermeer has made at least six paintings with a wall-map in the background. Three times it is the map of merely the province of Holland (Soldier with smiling girl; Young woman with letter; The geographer), once the map of Europe (Lute playing girl) and only in one other case (Young woman at the window) there is a vague wall-map suggesting the seventeen provinces. (Simon Schama, 1987, *The Embarrassment of Riches: an Interpretation of Dutch Culture in the Golden Age.*)

[32] Peter J.A.N. Rietbergen, 1992, Beeld en zelfbeeld. 'Nederlandse identiteit' in politieke structuur en politieke cultuur tijdens de Republiek [Image and self-image. 'Dutch identity' in political structure and political culture during the Republic].

[33] In twelve cities the estimated literacy level of males before the year 1600 was already well above the 50%. A.M. van der Woude, 'De alfabetisering', pp. 257-264.

the most univocal voice that reached all inhabitants was that of the Bible and the local minister. Clergymen were essential in distributing news about national events and interpreting them if necessary as analogous to what happened with the ancient tribes of Israel.

Simon Schama takes great pains to elaborate what he calls the Hebraic self-image of the Dutch.[34] The Old Testament provided a singular and most appropriate discourse about a tiny people guided to the Promised Land by divine providence and against overwhelming power. The covenant idea (Box 3.1) permeates the Old Testament, but it was particularly the story of the book Exodus, in which Moses guides his people through the Red Sea and in which God establishes his covenant with the Jews by giving the Ten Commandments, that was extensively represented in texts and in illustrations.

> And the LORD said unto Moses, Write thou these words: for after the tenor of these words I have made a covenant with thee and with Israel. And he was there with the LORD for forty days and forty nights; he did neither eat bread, nor drink water. And he wrote upon the tables the words of the covenant, the Ten Commandments. (Exodus 34:27-28).

The Dutch national anthem *Wilhelmus*, written somewhere between 1568 and 1572[35], is a peculiar monologue or rather a prayer from the great leader of the resistance against the Spanish, William of Orange. It tarts as an apology for his disobedience to the king of Spain ('To the King of Spain, I have given a lifelong loyalty'), but William had to follow the voice of conscience or God. Lost battles (Maastricht is mentioned) have 'scattered' the people, but God's guidance will award those who choose His way (see the verse at the beginning of this chapter). From the perspective of 1570 the political future certainly did not look rosy and bright and we can imagine that people could use some heartening thoughts while facing the formidable opponent of the Spanish Empire. It very soon became a popular song that allegedly could even chase

---

[34] Simon Schama, 1988, *The Embarrassment of Riches*.
[35] As a popular song it was used throughout the centuries at official ceremonies, but it was declared the official national anthem only in 1932. The 'national anthem' was a product of 19th century nationalism and the first Dutch national anthem was chosen in the early 19th century after a contest among contemporary poets and composers. Much later at the end of the century the winning text was considered too dull and people were looking for a connection with the dramatic events of the country's early history.

away Spanish army units – as once happened when a huntsman in the woods blew the melody on his horn.

> **Box 3.1. Covenant**
>
> The word covenant occurs 272 times in the Bible; of those instances, only 18 appear in the New Testament. In 1534 the Swiss theologian Heinrich Bullinger (1504-1575) published *The one and eternal testament or covenant of God*. For forty-four years Bullinger remained leader of the Reformed church of Zurich and his reformed theology was strongly permeated by the idea that the social order was founded on covenants, first of all the covenant of God with men like Adam, Noah and Abraham. It pictured the human world not as the appendage of a rigid cosmic order but as depending on deliberate agreement. The idea of covenant is central to the political concept of federalism (the word 'federal' derives from the Latin *foedus*, meaning covenant). Unlike Bodin's concept of sovereignty as authority exclusively reserved to the ruler, the federal principle assumes checks and balances and the creation of human communities based on what we may call a symbiotic order. The latter term occurs literally in the work of another writer in the Reformed tradition, Johannes Althusius (c 1557-1638), who lived in several places in the Rhineland and Switzerland and ended as chief-magistrate of the city-state of Emden near the Dutch border and the North Sea. In his book *Politics* (1603), Althusius describes the world order as an expression of decisions attuned to the building of communities that fulfil human life and destiny rather than an expression of natural law. Society is built from different levels of symbiotic groups: families, voluntary associations (guilds), cities, provinces and the commonwealth. Each new level is created by means of a covenant among representatives of the lower level groups. In contrast with the Bodinian concept of central sovereignty this organization is characterized by the division of powers; each level checks the politics of the lower and higher levels on the basis of its covenants. In Althusius' politics there is no real sovereignty of the people but at least a mutual check of authorities that makes resisting tyrants and removing magistrates from office legitimate. It also allowed for resistance against 'foreign' invaders by inhabitants of a territory. This had recently happened in the Dutch revolution (1579) and would be repeated a hundred years later in the English revolution of 1688 and again in about the same time interval in the American Revolution (1776-1783).

However widely the Bible was read and used to interpret politics, the linking of daily life and religious phenomena had a strongly secular overtone in the Dutch Republic. Despite the Republic being a hotbed of Calvinist movements and fundamentalist factions, Dutch Republican leaders maintained a strongly liberal view in politics. This view was rooted in the merchant class and was cherished by political refugees who had escaped intellectual suppression in the Spanish Netherlands or in France. They found much utility in the image of a travelling people, on their way to the Promised Land amidst hostile forces (both of social and natural origin), and in the story of laws made to produce order in a society where religious fanaticism could play havoc. Had not Moses, when he descended from the mountain with the tables on which the Ten Commandments were written, found a people that indulged in false religious practices by worshipping a golden calf? According to Simon Schama we should interpret the painting of Moses with the Ten Commandments which Ferdinand Bol painted for the new Amsterdam Town Hall precisely in this way: the regents of the town warned the Calvinists not to interfere in matters of city politics.[36] This notion fit neatly with the Dutch philosopher Erasmus's humanistic view that "towns ruled by priests stagnate and fall into ruins".[37]

The mere multiplicity of political actors, regents, cities, states, princes and clergymen forced all players to engage in mutual arrangements lest they end up in civil war or political instability and inevitably elicit foreign interference. But it left territorial identification ambiguous for a long time. Maps of the seventeen provinces were left curiously empty at the southern edge because nobody knew where the boundary should be drawn and what cities to include in the Republic.[38] The image of a wandering people and the openness of covenants to enlargement pushed the territorial question in the background. Apart from delimiting national identity in an unstable international world, the discourse of the Old Testament provided a useful language for discussing political relations in a dynamic economy.

Hebraic discourse was not unique to the Dutch Republic but emerged in one form or the other in all places where the Reformation became firmly embedded in the 16th century. Small and dispersed communities of believers felt heartened by stories about the power granted to the weak by the faith.

---

[36] But see note 31 on Schama's enraptured interpretations.
[37] See page 61.
[38] H.A.M. Van der Heijden, 1998, *Oude kaarten der Nederlanden*, p. 114.

England's Protestants propagated the idea that the English, due to God's special care, had escaped dark political forces, particularly those associated with Catholicism and France. As told in the stories of the wandering people of Israel or the traveller in Bunyan's edifying poem *The Pilgrim's Progress*, salvation was not found without having been exposed to hardships and evil powers.[39]

In 1653 Oliver Cromwell called Exodus "the only parallel of God's dealing with us that I know".[40] This judgment coincided with attempts to restore the Jews, expelled in 1291 by Edward I, to England. The background of English philosemitism as it rose in the 17th century was complex. A religious strand of thinking embraced the idea that the religious purity as developed in England would convince the Jews that in this country the Biblical prophecy had come true, which would convert them and bring about the arrival of Christ's millennium. A more pragmatic argument was that Jews were connected to trade and that trade is essential for liberty, *ergo* Jews fit the English national values. Other voices praised Jewish knowledge, clearly flowing from a wandering and cosmopolitan way of life. All these arguments assumed the congruence between the Jewish world and English national interest.

One might wonder why this infatuation with a foreign people or foreign roots arose in a country that already knew a long process of territorial appropriation and integration culminating for the first time in the Tudor creation of the English 'state' under Elizabeth I (1558). There are two possible answers to this question, one referring to the anomalous nature of the apparent English *state* and the other deriving its argument from the reluctant acceptance of a *British* identity between 1603 and 1707.

In England, the monarchy never emerged from the medieval Christian Empire with the great clarion call of the modern territorial state like it sounded elsewhere. Already in the fourteenth century the Commons functioned as a countervailing power in the English body politic.[41] Whereas kings elsewhere enhanced their legitimacy by adopting a divine status, in England an early (1401) Speaker of the Commons could compare the body politic of the English realm with the Holy Trinity: the King, the Lords *and* the Com-

---

[39] Linda Colley, 1992, *Britons. Forging the nation 1707-1837*, pp. 18-32.
[40] John K. Hale, 1996, 'England as Israel in Milton's writings'. The author focuses on the absence of 'the only parallel' (Cromwell) in Milton's writings.
[41] Ernst Kantorowicz, 1857, *The King's two Bodies*, p. 227.

mons. The English sense of unity was based on a tradition of common laws and institutions integrating, rather than centralizing, the territory like a state does.[42] Additionally, this weakness of the state was not compensated by a strong ethnic definition of England. Both kings and those who opposed the crown ('patriots') saw themselves as actors within a larger, extra-national framework.[43] During the entire 16th century this framework still surpassed the British Isles; it was a European universe – albeit either a Protestant or Catholic one.

After Elizabeth's death in 1603 her Scottish nephew, King James VI, was the only legitimate successor to the throne. His reign (as James I) started a regnal union of Scotland and England but did not yet integrate the governments or parliaments of both countries. Already before this event the prospect of British unity had evoked enthusiasm in some circles, particularly in Scotland. At mid-century the prospective union was hailed in apocalyptic terms as performing a central role within the great sacred drama of the Scriptures. For some the message of James' accession to the British throne was such that the apocalyptic drama would now enter upon its final acts. Such terminology in which the Scriptures were endowed with geopolitical meaning was something new. It implied that the image of the Antichrist changed from an individual to that of an earthly power. The shift was already perceptible in 1539 when the Earl of Glencairn denounced the Catholic clergy as '*Monsters with the Beast's marke*'.[44] However, English apocalyptic expectations were usually incited by English institutional reality rather than by a glorious Britain.

As literary historian Weinbrot asserts, Britons, in spite of continuous integration in the eighteenth century, remained aware of diverse ethnic strings and foreign influences running through their history. Scholars and writers consequently interpreted this diversity as a typically British asset in a competitive and changing world.[45] Only backward countries, resistant to change, so the argument ran, would preserve their original homogeneous nature. Such a perspective was of course receptive to the creative 'discovery' of ancient

---

[42] Kenneth Dyson, 1980, *The State Tradition in Western Europe. A Study of an Idea and Institution*, pp. 36-44. Of course, there was a more organic centralising force in the shape of an expanding government in London!

[43] Arthur Williamson, 1999, 'Patterns of British identity. Britain and its rivals in the sixteenth and seventeenth centuries', p. 141.

[44] Ibid., p. 145.

[45] Howard D. Weinbrot, 1993, *Britannia's Issue. The Rise of British literature from Dryden to Ossian*, pp. 557-571.

(Celtic *and* Jewish) inputs in the British (or English) people. Together with the idea of a people united by the value of freedom rather than by coercion in an authoritarian state, this bolstered both the analogy of the Hebraic narratives and sympathy for contemporary Jews.

Territorialization of the boundary

In November 1659 a French company of nobles, including the highest diplomat and royal secretary cardinal Mazarin, prepared to set foot on the floating bridge that led to a small Pheasant Island in the Bidassoa river. At the same time a similar company, led by the Spanish chief diplomat Don Luis de Haro, approached the island from the opposite bank of the river where they also encountered a floating bridge. Both companies entered the pavilion, erected on the island for peace talks, through their own entrance. The outcome of these talks would become known as the Peace of the Pyrenees, a somewhat belated reverberation of the Westphalian Peace (1648) where the problems between France and Spain had been left out because France hoped to gain more from bilateral talks. The entire setting of the talks reveals the changed meaning of the boundary and the presence of sovereign states rather than an empire dealing with barbarians. Everything is carefully arranged to express symmetry and equality. The reality is that one side, France, is the most powerful and can expect to cash in the most, but the ritual is one of equality and mutual recognition. The Bidassoa flows through the lowlands at the west side of the Pyrenees Mountains and was already accepted as the boundary between France and Spain. Both delegations decided to declare Pheasant Island neutral territory for the time of the talks. The Pyrenees Mountains had never been contested as a rough boundary between the two countries, but the precise course of the borderline needed to be fixed. This was seen as merely a technical problem and delegated to a commission of jurists. There were weightier political questions under discussion, such as the Spanish Roussillon near the Spanish border – its main city Perpignan was recently conquered by the French – and the intended marriage between Louis XIV and Maria-Theresa, a German niece of the Spanish king Philip IV. The French had decided beforehand that they would never sign the peace without acquiring Roussillon and Artois (in the north of France). On both territorial matters the Spanish conceded, but the rashly referred task of delineating the boundary appeared to run up against formidable obstacles. First a simple cue for draw-

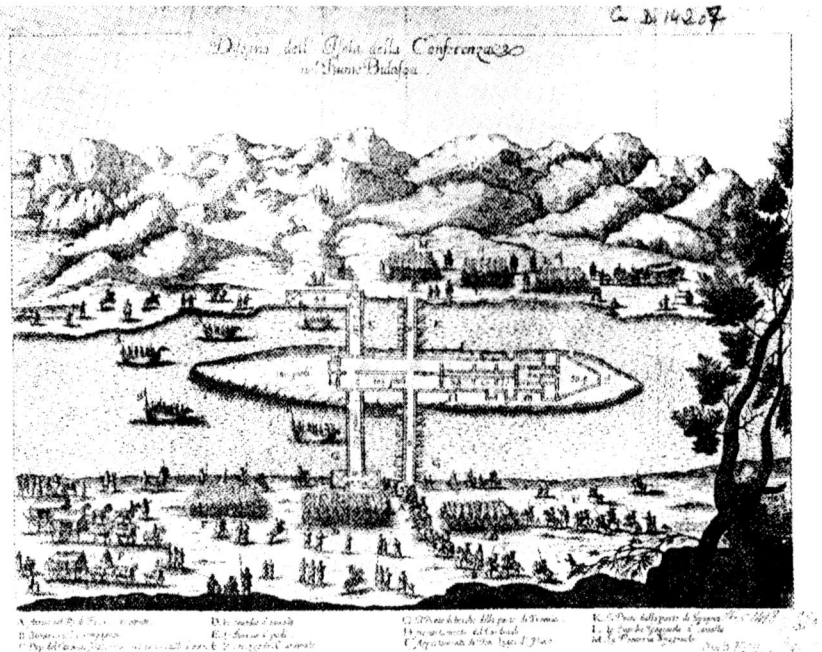

Figure 3.2. Pheasant island in the Bidassoa river. The scene of the Peace of the Pyrenees talks (Source: Peter Sahlins, 1989)

ing a boundary as the line that connects the summits of a mountain range is not so easy to apply, let alone that it may lead to results that are not intended at peace talks. Mountain ranges often have several parallel ridges and switching from one to the other may cause indeterminate results. Where mountains levelled off at the end of a range, the guideline to follow the highest summits could even cause the borderline to bend wide away from the intended direction. One such 'logical' course from a Pyrenees crest over the contiguous Corbières hills would even give the Roussillon to Spain. But the most confusing problems occurred in the human sphere and concerned the fact that land property, religious institutions and even the military were not equipped to accommodate the phenomenon of a state boundary. Some of these problems complicated the fixing of the boundary and most of them influenced the practical acceptance of the boundary. The latter took many decades and required essential political and social adaptations described by Peter Sahlins[46] – who wrote a fine history of this process in the wake of the Pyrenees Treaty

---

[46] Peter Sahlins, 1989, *Boundaries: The Making of France and Spain in the Pyrenees.*

Chapter 3. A NEW JERUSALEM

– under the title of 'territorialization of the boundary'. Sahlins selected a small region for his research, the Cerdagne (Fr) (Cerdanya (Sp.)), a valley, almost a plateau of 20 kilometers, between two northeast-southwest running ridges. It is one of those areas where the borderline had to switch from one ridge to another and where the plotters of the line lost solid ground. The Cerdagne as such was located on a strategically important connection, an elongated 'glen' quite open to Perpignan and its surroundings (the Conflent). In the middle of Cerdagne lies the urban settlement Puigcerda, and in the 17th century the valley itself was speckled with sixty-odd small hamlets or villages. Control over the valley was important from a military point of view because it provided food and revenues (taxes!) to a garrison and because of what we may call the military mode of production at the time. The idea of accepting the high ridges of a mountain as a borderline made no sense from a military point of view. Soldiers needed to occupy the high ground with fortifications and look out over lowlands in order to nip in the bud each deployment of offensive equipment by a hostile army as far as one could see. So the traditional practice of establishing military frontiers or buffer zones still prevailed and drawing a line was anathema to this practice. Faced with the inevitability of a divided Cerdagne, each side tried to retain as much territory as possible at the foot of the ridge in order to facilitate army movements (hampered by transversal gorges) in a longitudinal direction.

The resulting situation left both sides perplexed; the militaries were at a loss as to how to handle the new conditions. For decades after the borderline was fixed, one remained inclined to perceive the situation as one of dominance over the valley. After Puigcerda had been bequeathed to Spain the Spanish army erected fortifications in the town, which immediately incited the French garrison in Mont-Louis to send an expedition in order to demolish these structures. Irrespective of what was decided in the treaty, the entire Cerdagne was alternately occupied by either French or Spanish troops during the period from 1660 to 1720. In practice, no borderline functioned. But in 1720 something happened that catalysed the acceptance of border guarding as a natural practice; so to say the mental internalization of the boundary had occurred. From Marseille the bad news reached the Cerdagne that the plague had broken out. This was the kind of event in which one could count on established practices of territorial closing: the *cordon sanitaire*. In the Cerdagne this usually and most easily was implemented by closing the mountain passes. The Spanish, who happened to occupy the valley at that moment, did just that. But whereas military procedures had violated the Peace Treaty time and again, this civil measure seemed to provoke the adversary in a more fundamental sense. French officials protested against this violation of the treaty and demanded that the checkpoints transfer to the official borderline. Some sheds were hastily erected in the valley for checking personal movements. They were manned with officers of both sides and henceforth this procedure was accepted as valid boundary practice.

Just how difficult the acceptance of a negotiated borderline between two sovereign states in the 17th century still was is also illustrated by figure 3.3. The title of the map reveals how the negotiators proceeded. They took a complete list of villages in the Cerdagne that would have to be more or less split in two. One village would go to one side, the next one to the other, of course taking into account the rough geographical positions. Here the 33 villages that went to France are shown. The map also shows another curious detail: mountains mark the boundary that crosses the valley from north to south whereas no such landmark exists in reality, as the 3D map of the same area (figure 3.4) clearly indicates. Apparently, there was a strong need to naturalize the territorial state. This is in accordance with the idea of sovereignty as a transcendent truth, something given by God or nature and not completely deducible from human decisions.

Even the idea of starting from a list of villages reflects a way of thinking that better suited the world of personal relationships than the new one of territorial sovereignty. The juridical definition of a village was not a territory but a list as well: the enumeration of a number of people or names. Nothing guaranteed that their land would form a nice and contiguous area around the village nucleus. This means that the decision to bequeath a village to one state does not solve ambiguity in establishing the precise course of the boundary. No cartographic decision could avoid the conundrum of land property being located on both sides of the boundary. The personalized approach to spatial configurations even meant that not all the territory in the valley could be accounted for. Deserted villages were, according to this definition, non-existent, white spots on the map. Only by historically tracing who had lived there and where those people had gone was such a village legally assigned to 'the territory' of another village – but this would make a political cartographer hardly less desperate. One of the embarrassing consequences of this negotiation game with villages was that the Spanish side, after the agreement claimed an exceptional status for the settlement Llivia because it was juridical-

Chapter 3. A NEW JERUSALEM

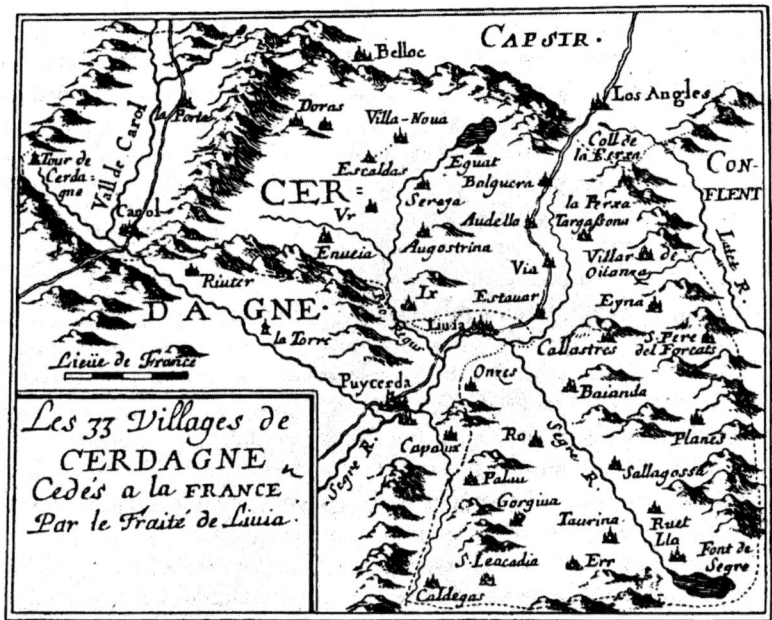

Figure 3.3. The 33 villages ceded to France in the Treaty of LLivia.

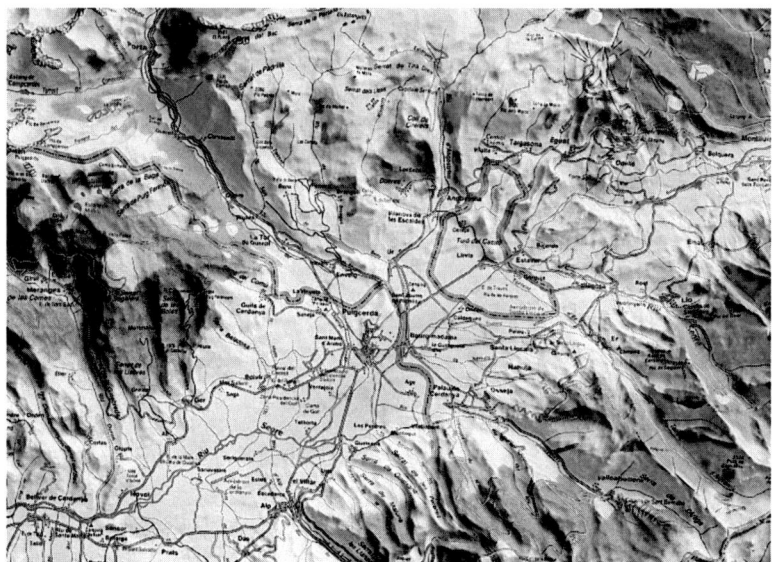

Figure 3.4. Relief map of area from figure 3.3. Mind the absence of the North-South mountain ridge of figure 3.3 along the separation line.

ly not a village but a town. This explains the current existence of a Spanish exclave with that name.

Just as the military could not immediately adapt to a boundary once it was established, civil society and authorities remained for a long time untouched by the new dividing line. The Spanish king Philip IV, bearer of the title 'Duke of Roussillon', initially refused to recognize why he could not maintain his title. Seignorial rights of convents on Spanish territory kept being enforced in French Cerdagne. Such bodies could not only collect tenths or annuities but also administer justice. The French province of Roussillon maintained the right to demand customs so that people travelling between Spain and France had to pay customs twice. Apparently, the process of territorial sovereignty, which Jean Bodin had characterized hundred years before as something quite obvious, was still far from achieved.

Yet among European states the French state was the most advanced in centralising all civil liabilities like taxes, justice, social needs like security and health provision and all types of authority (even the Church). There was already in the 16th century a clear urge toward a juridical framing of French citizenship, which assigned rights to the French (like the inalienability of property) that did not apply to the foreigner on French soil.[47] This did not mean that national unity had also entered the feelings and perceptions of the people, but it was a step in that direction. A French inhabitant of the Cerdagne for a long time felt more related to the people of Spanish Cerdanya than to a French citizen outside the Pyrenees. Barcelona rather than Paris was the obvious place to move for those who pursued a non-local career. Only in the course of the 19th century did this territorial identification change under the influence of national history writing, education and military conscription. A decisive moment was the Franco-Prussian war of 1870-71, when people from the most remote places gathered and died for a common goal. The war also revealed how futile any military presence at the boundary was in defending a country. Military maps of the boundary regions did not even represent the topography of the neighbouring state anymore; all land across the boundary had become a white space. This was an odd procedure for armies that, in order to win a battle, cannot simply stop where the map ends at the state boundary. However, it rightly expressed that the most relevant military reconnaissance had moved to the places where war was actually planned: the capital cities.

---

[47] Charlotte C. Wells, 1995.

Chapter 3. A NEW JERUSALEM

Early-modern territorial principles and moral government

From the era of the Italian city-states to the American and French Revolutions, 'self'-determination or territorial sovereignty was never an easy matter, neither in practice nor in representation. In the 16th century, self-determination and self-government, however incomplete, had to be implemented in a world where such notions did not yet exist. Liberating a community from imperial or dynastic chains only started the problem of who should represent the 'public interest' (identity) and how to engage with an international order (closure). While the idea of territorial sovereignty and the state boundary as a crucial discontinuity in authority gained acceptance, the state's representatives still faced uncharted territory in two directions: inward looking they could not zero in on a hierarchic system of social control, and outward looking they could not find international rules of engagement. As Erik Ringmar has noted, state leaders had to invent a role for themselves as international actors. Europe had suddenly become a stage on which they performed as in a play (a 'Game of Thrones' in current jargon).[48] This was a clear break with imperial territoriality in which an empire's identity and the role of its lords and knights was beyond discussion.

In the domestic sphere, the idea that people were linked to a territorial commonwealth rather than to a lord frightened the landed aristocracy who saw a contradiction between their established property rights and the process of being turned into common territorial 'subjects'. In the German principalities of the 16th century, princes or 'landgraves', who were in some cases legitimated by their role as protectors of the (Protestant) belief, pursued the new 'closure' ideal. Their ambition, however, to assume absolute power was blocked by the domestic estates (prelates, towns, knights) who could appeal to the higher German imperial authorities (the Holy Roman Empire of the German Nation) when such vested interests were endangered. Such struggles over authority nevertheless strengthened the idea of territorial identity by stimulating a discourse on the responsibility for and love of the homeland. In early 16th century Hesse, an association of knights seeking to augment their influence on the affairs of the landgrave was not considered unusual, let alone illegal. However, in 1647 the Hessian successor Amélie called meetings of the

---

[48] Erik Ringmar, 1996, *Identity, Interest and Action. A Cultural Explanation of Sweden's Intervention in the Thirty Years War*, p. 155.

estates to discuss their response to new taxes 'treason'.[49] This reaction was in conformity with the Bodinian idea of territorial sovereignty resting in the Crown. In the resulting lawsuit at the Imperial Chamber Courts the estates argued that they also represented the country and could meet without consent of the prince. They actually appealed to ideas similar to those propagated by Althusius (Box 3.1): that there is a broader responsibility for the welfare of the fatherland (commonwealth) than that of the monarch alone. Paradoxically, the conflict about interests produced a discourse that reinforced territoriality, although it did not immediately evoke the type of territorial closure existing in France or England. This was because many actors had an interest in maintaining the two levels of jurisdiction: state and German Empire!

The problem was that the new territorial order remained highly indeterminate. *Closure-identity* was the obvious challenge in politics but could be manipulated by rulers like the Electors of Brandenburg-Prussia who promoted migration of Calvinist groups to newly conquered areas like Silesia.[50] The unfinished nature of the assertive early-modern polities further complicated the picture. This applies to the Dutch Republic with its provisional southern frontier and to pre-Revolutionary America with its nameless areas to the west. Even the much-praised insularity of Britain did not immediately elicit a politically unified territory. The same pertains to the Italian communes whose authority did not stop at the city walls. In the countryside (*contado*) around the city, the gradual take-over of seigneurial rights by the communes continued like it had happened previously in the urban area proper, although with stiffer resistance from established powers.[51] In the *contado* the solution often was some form of power sharing or the concluding of feudal-type contracts with the inhabitants of villages. The result was a patchy, non-concentric extension of urban influence with various levels of intensity.

While the ideal of territorial separation reigned supreme in Early-Modern Europe, boundaries rarely separated neighbours that went through the same modern-state dynamics. England, Sweden, the Dutch Republic, France and New England bordered on dynastic-imperial (Habsburg Low Countries), semi-colonized (Ireland, Scotland), hybrid-imperial (Holy Roman

---

[49] Robert von Friedeburg, 2005, 'The making of patriots: love of fatherland and negotiating monarchy in seventeenth-century Germany'.
[50] Christopher Clark, 2007, *Iron Kingdom. The Rise and Downfall of Prussia 1600-1947*, pp. 115-144.
[51] Philip Jones, 1997, p. 361.

Empire) or natural areas (Pyrenees). Switzerland was a natural buffer area itself. This means that boundaries could long remain indeterminate and that local cross-border communication continued. Since ethnicity or ethnic geography was no valid criterion in establishing boundaries, the idea of a wandering people (Israel) on its way to a (transcendental) destination perfectly fit the situation. It also fit a territorial structure that was federal (tribal) in shape and unfinished, or federal because it was unfinished. Even the French kings 'made no effort whatsoever to overturn the customary laws of Normandy, Brittany or any other province'.[52]

The only available ideological background for attributing legitimacy to 'communities' was an embryonic frame of thought about communities of faith and covenants and the work of a few political writers who distanced themselves from Bodin's absolutism by basing themselves on the Italian city-states and citizenship.[53] The city also offered the most realistic model to capture the new civilian character of the state and this particularly fit France in the 16th century, notwithstanding its centralist tradition and philosophy. An explicit link was made with Italian medieval city-states, and through them with the Roman Empire. In 1569 the French constitutional law specialist Jean Bacquet commented on the French state as 'one great city after the fashion of the Roman Empire'. This strongly appealed to legal conceptions of citizenship and the rights of Frenchmen who stayed abroad for some time and, conversely, foreigners staying in France.[54] The increased mobility of people and the amassment of wealth elicited juridical questions concerning property and inheritance rights. Whereas the dynastic idea of considering people merely as subjects of a King was essentially a non-territorial principle that remained in full force for individuals who moved outside the Kingdom, the territorial and citizenship approach started to introduce such distinctions as

---

[52] James B. Collins, 1997, 'State building in Early-Modern Europe: the case of France', p. 622.
[53] Jean Bacquet and René Choppin in France (see Charlotte Wells, 1995, *Law and Citizenship in Early Modern France*) and Paul Buis (Busius) in the Dutch Republic (see Kossmann, 2000, p. 34).
[54] Charlotte C. Wells, 1995, *Law and Citizenship in Early Modern France*. It is typical for the age of Early Modernity that several conceptions of authority and territoriality were voiced at the same time and alternated in gaining acceptance as official doctrine. Unlike Jean Bacquet, Jean Bodin clung to the citizen's obedience to a single authority at the expense of citizens as a community. His conception foreshadowed the absolutism that would dominate France during the next two centuries.

the length of a person's stay in or outside the territory, the place of birth of his or her parents, etc. A citizen was someone who worked for the common good and who could consequently appeal for rights in the state. Here the city-metaphor was tacitly linked to religious connotations like the comparison that Erasmus made between city and monastery: all inhabitants share the same devotion to the common ideal.

The common good was inextricably bound up with land as an agricultural resource, something that conceptually fitted the new territorial preoccupation with boundaries, geometry and its application in the cadastre, particularly in France.[55] The bulk of tax revenue for the state was derived from landownership or agricultural production, but the introduction of a just system of taxation and the implementation of the cadastre remained a stumbling block well into the 18th century, not in the least because of stiff resistance from the nobility. The mindset of the early modern state was attuned to governability, but a logical practice of control often stopped short at discussions among representatives from different estates. The gap between the new spirit of transforming territory into a system of production manifested itself also in the practice of *cameralism* that was more typical of the German principalities.

The label 'cameralism' refers to an economic theory about the state that suggests principles and policies for optimizing its wealth, something like mercantilism but without the overwhelming emphasis on trade. The term derives from the German word '*Kammer*' (room), originally indicating merely the princely treasury and later on evolving into a designation for the entire financial administration of a principality.[56] German cameralists of the 18th century were extraordinarily impressed by the English factory system. They applied the mechanical metaphor of mills and cogwheels to the running of estates and subsequently to the functioning of the state as a whole. The three key elements were fiscal control, exploitation of resources and economic regulation. The bureaucratic ideal of a 'police state', originally meaning a state in which the government directs the economy, is clear enough in the work of the cameralists, but most historical studies have concentrated on their ideals without checking the practice. The practices of a writer who is widely recog-

---

[55] Charles C. Maier, 2016, *Once Within Boundaries. Territories of Power, Wealth and Belonging since 1500*, Ch. 3.
[56] Michael Jackson, 2005, 'The eighteenth century antecedents of bureaucracy, the Cameralists'.

nized as *the* personification of cameralism, Johann von Justi (1717-1771), were actually an unmitigated disaster.[57]

The state's pursuance of order revealed the strategic role of ordinary people in a way other than as producers: a central state aspiring to things other than tax collection (which was not without implementation problems either) badly needed the knowledge and agreement of local people. There was no solid bureaucracy in the 16th century reaching from the sovereign ruler to the level of the village and the street. The state, on its way from a static entity to an active organization, was first dependent on the local level where a culture of governance had to germinate that could favourably intersect with regulation by the Crown. Many of the activities that we currently consider as typical for state care were privatized. 'Prisons for example, were private businesses whose keepers bid for leases and expected to make a profit for the fees that prisoners paid for accommodation and food'.[58] As penetrating studies of early modern England have shown, state regulation pivoted on local caretakers of the middling sort: petty constables, churchwardens, yeomen and overseers.[59]

People in late Elizabethan and early Stuart England were well aware of the changing times which unfortunately seemed to boil down to social disorder, riots, mutual violence and property crime. Since economic interpretations of social change were still unknown, social problems were traced to vice. However, instead of merely emphasizing repentance and pious rituals as was customary in an earlier period, a system of redressing the social order emerged in which (common) law and poor law sowed the seeds of a true public policy practice. Christian ethics shifted from demanding the right individual attitude toward developing a series of institutions and punishments.

---

[57] Andre Wakefield, 2009, *The Disordered Police State: German Cameralism as Science and Practice*. Chicago: In Wakefield's detailed reconstruction of the events, however, von Justi, who had impressed many by his writings, was a completely impractical scholar who tried to boost iron production without taking into account such factors as the carrying capacity of woods and labour management. His commission in 1765 to build a new steel and iron plate works east of the Oder (in current Poland) got so much out of hand that von Justi was relieved of his commission in 1767. Subsequent charges brought him in jail where he died as a blind man in 1771.
[58] Jeremy Black, 2001, *Eighteenth-Century Britain, 1688-1783*.
[59] Steve Hindle, 2000, *The State and Social Change in Early Modern England, 1550-1640*.

> A charge of 1623 empowered parish officers to report any idle labourer…; to ascertain every week which of the poor had work for the next week and to supply materials on which they could be employed; to ensure that pauper children be taught knitting and spinning; and to make twice-weekly search for suspected nightwalkers and pilfered goods. Most tellingly of all, however, overseers were not only to punish but also to withhold poor relief from all beggars and pilferers. The money thus saved would be used to reward those inhabitants who informed on pauper delinquency, at the rate of sixpence a time.[60]

Paternalistic in nature, these policies had to deal with the difficulty associated both with the resistance of people targeted by such policies as well as those charged with new duties. In the end, a political community was generated in which local middlemen knew how to use new responsibilities to their own advantage and in which others increasingly had recourse to litigation when they felt victimized by power abuse. The system of law tied the local level to the Crown, but its courts were reticent to lay down strict rules of implementation that everywhere applied (applying top-down state governance). The state was rather welled up from below by a cultural change in which claims to authority were continuously held against the light of public judgment and the law. Similarly, in the Dutch Republic some type of governance was inspired at the local level by the 'struggle against the water', but it was never a concern of the state (or an incitement to nation building) until the 19th century. The historic role attributed to this struggle was rather a product of 'invented tradition', a nationalist practice from the high-modern period.[61]

Early modern governance of the domestic scene aimed at 'moral order'. It had religious roots but the centre of governance had shifted from the church to the public domain. The parish church, the traditional meeting place for dignitaries of the village, continued to be the place for official meetings but leaders now met there under the royal arms rather than under the cross, or at least under both.[62] Governance was not something unrolled over a state territory like a carpet with a fixed pattern. Early modern governance had germinated in the local community and was neither necessarily functional for the state as an economic 'growth machine', nor an expression of the wish to provide social justice. Such aspirations prefigure a new (high-modern) age, but

---

[60] Ibid., p. 174.
[61] Hans Knippenberg, 1997.
[62] Steve Hindle, 2000, p. 229.

they were already expressed in the 18th century by thinkers such as Adam Smith (1723-1790). Adam Smith is often erroneously identified with the *laissez-faire* state and the 'invisible hand' that would regulate the economy (but that according to Marxists ultimately enables capitalist to exploit the workers). In his *The Theory of Moral Sentiments* (1759) and *Wealth of Nations* (1776) the aim was rather the opposite. His work was an attack on 'the corporate, municipal and parochial institutions, and in particular the uncertain jurisprudence in which these institutions flourished'.[63] To put it in other words: such local instruments of governance often turned out to protect established interests and to limit the possibility of others taking advantage of commercial opportunities. Institutions like apprenticeship (at that time recommended for their support of moral order in society) actually kept rewards high for workers and masters by limiting access to the labour market. As instruments of monopoly they denied the principle of equal opportunity advocated by Smith. In order to restore equality, he was even prepared to defend regulation of minimum wages.

The word 'sentiment' in Smith's *The Theory of Moral Sentiments* already indicates his rejection of a moral order that is based on established institutions or religion. Moral sentiments are the motives and expectations that are aroused in the daily interactions between individuals – for example the reliability of a commercial partner, freedom of choice or the wish to be recompensed for labour. This immanent or liberal foundation of rules for economic behaviour (meaning rules revealed in the human actions themselves) lacks the certainty that appertains to religious commandments or the laws of an absolute monarch. The absence of authority (just like in democracy) evoked what Smith called 'the very suspicion of a fatherless world'[64], words that stimulated philosophical reflections of enlightened thinkers but that eventually contributed to the territorial shock of the coming High-Modern age.

The early-modern impact of closure and identity

The metaphoric steps from monastery to city (Erasmus) and from city to state (Jean Bacquet) that were made within a few decades in 16th century po-

---

[63] Emma Rothschild, 2001, *Economic Sentiments. Adam Smith, Condorcet and the Enlightenment*, p. 88.
[64] Ibid., p. 229.

litical writing reveal the territorial challenge in the Early-Modern age. It indicated the new importance of people (citizens-monks) in establishing the substance or identity of a political territory (state-monastery). It also emphasized the inducement of right attitudes (*governance*) in citizens in order to produce what is nicely labelled 'New Jerusalem'. Since this implementation of *governance* was slowly progressing at the level of the central state – at least compared to the later (High) Modern state – the impact in the sphere of *closure* and *identity* was most substantial (table 3.1).[65]

Table 3.1 Early-modern territoriality

Dimension	Active/passive	(Un)boundedness
Closure Structure of authority. Who are (un)affected by dominant authority?	Shifting internal and external boundaries (active)	Territorial sovereignty Diplomacy Elementary citizenship
Governance How is a territory / group engaged in turning it into a resource?	Dependency on local caretakers (passive)	Collecting revenues (taxation) Keeping social and moral order Property law
Identity How do people attribute *meaning* to power in (territorial) assemblages?	Searching for territorial distinctiveness (active)	State<>religion Becoming citizen

In a world where boundaries where vague, transitory, or mainly used for special purposes like the parishes and the bishoprics of the Church, the rise of a central authority obviously needed some moral legitimation. The image of a wandering people escaping the dark forces of the world guided by divine instruction was helpful in some cases. Outside the reach of the Reformation (France), the state itself or the monarchy was portrayed as specially favoured by God. Such moral legitimation of closure did not provide any clues to the nature of the international system. Here we witness the transition from the imperial to the state mode of territoriality with the consequence that the international scene became a game (or theatre) with uncertain rules.

Although the rise of central authority entailed a new pattern of (social) governance, the territorial shock was rather unleashed at the interface of closure and identity, as Dante's case already illustrated in the preceding chapter. These territorial aspects, which showed a high momentum like the hard-

---

[65] As Michael Roberts remarks in his study of Sweden under Gustavus Adolphus the clergy could be 'great public servants. In education, in the primitive social services of that age, they were the natural instruments of governmental politics.' Michael Roberts, 1953 (1958), *Gustavus Adolphus. A History of Sweden 1611-1632*, p. 369.

ening and shifting of boundaries and the rise of an ultimate secular authority, were important for one's self-image in the world. They were shocking because they clashed with the medieval conception of multiple authorities that only obeyed the single representative of God. The evolving system of *governance* did not immediately penetrate the level of common people. Armies were still recruited from professionals or mercenaries and daily life was more affected by the activities of the middling sort of people or estate owners. This aspect of territoriality would become energized in the next epoch.

Faced with the unremitting suspense about internal and external political relations, the biblical narrative was a godsend. Indeed, Lutheran theology denied the possibility of encountering God in worldly affairs, but this was mainly useful in legitimating the existence of sovereign rulers other than the pope or a single Christian emperor. Jerusalem and the Exodus narrative provided metaphors that were indispensable in building collective identities in new states. Moreover, apocalyptic narratives utilized biblical predictions about what would actually happen on Earth, in a vague geopolitical sense. The period of Early Modernity shows how religion is reinterpreted to accommodate a new territorial world in which closure and identity are on the move. Its result was a hardening of the political space that finally could do without the narrative of a chosen people.

When the spheres of domestic politics and international relations were codified by law, and states could trust the loyalty of their populations, a more secular discourse started to dominate politics. How a religious discourse on the nation is cut short when a polity faces its internal consolidation is aptly illustrated by the history of the American Great Seal. When the American Congress agreed to the Declaration of Independence (1776) it appointed a Committee to devise a seal for the new union. Two members of the Committee, Benjamin Franklin and Thomas Jefferson proposed Old Testament themes. Franklin's design, under the motto 'Rebellion to tyrants is obedience to God', showed:

> Moses standing on the shore and extending his hand over the Sea, thereby causing the same to overwhelm Pharaoh, who is sitting in an open Chariot, a Crown on his head and Sword in his Hand, Rays from the Pillar of Fire in the Clouds, reaching to Moses, to express he acts by Command to the Deity (Figure 3.5).[66]

---

[66] Conor Cruise O'Brien, 1998, *God Land. Reflections on Religion and Nationalism*, p. 61.

Figure 3.5. Benjamin Franklin's design for the Great Seal

In the end none of the Old Testament themes were chosen, and after some years the Congress adopted the Great Seal in its current form. The eagle suggests continuity with the world of ancient Rome rather than with that of Israel.[67] This secular geopolitics was echoed in the Federalist Papers as a theory of domestic governance that attempts to control human behaviour by the design of (territorial) government rather than by methods that directly impact the human individual. Freedom has become the moral imperative in America, although one could argue that it is also still considered a divine commandment.

---

[67] But the reverse shows a pyramid and an eye. '*The pyramid signifies Strength and Duration: The Eye over it & the Motto [ANNUIT COEPTIS, Providence has favored our undertakings] allude to the many signal interpositions of providence in favour of the American cause*', as the designer Charles Thomson noted in 1782. Underneath these symbols the words 'NOVUS ORDO SECLORUM, A New Order of the Ages' appear. See also www.greatseal.com

Chapter 4
THE VERTIGO OF PUBLIC SPACE (HIGH MODERNITY 1815-1980)

> …what previously in long winter nights
> the old book had told me
> now resounded loudly in the forest's splendour,
> I heard it ring out brightly:
> in the forest at Vogelweide
> I also learnt how to sing.[1]
>
> (**Richard Wagner**, *Die Meistersinger von Nürnberg*, 1867)

> The current condition of nations is a result of the amassment of creations, improvements, refinements and efforts of all generations that have lived before us; these constitute the cultural capital of the currently living, and every single nation is only productive insofar they take in and extend by their own attainments, the achievements of former generations (…)
>
> (**Friedrich List**, *Das nationale System der Politischen Ökonomie*, 1841, p. 228)

Songs and territory

Richard Wagner's opera *'The mastersingers of Nuremberg'* moves the audience mentally to a place in Bavaria where a song festival, to use a modern designation, is to be held. But the scene is not modern at all; it portrays a medieval type of social setting with masters and apprentices, a guild society. In contrast to today's song festivals, the contestants are exclusively male. Actually, the event is supposed to have happened somewhere in the middle of the 16th century, the period when 'modern' Germany was born. The contest is held with the intention of giving candidates the opportunity to acquire the title of master. The master contest is a ritual that demands the right application of rules about the number of stanzas, the theme and the return of the melody and in what manner and so on. A stranger, the knight Walther von Stolzing,

---

[1] …was einst im langer Winternacht / das alte Buch mir kund gemacht, / das schallte laut im Waldes Pracht, / dass hört ich hell erklingen: / im Wald dort auf der Vogelweid', / da lernt'ich auch das Singen. English translation from Peter Branscombe. The song and melody are somewhat queer which aims to induce the same bewilderment in the real opera audience as the audience in the Nürnberg song contest is supposed to feel.

announces that he would like to participate, but he is clearly ignorant of the rules. Asked about his master, he refers to the minstrel Walther von der Vogelweide, who actually died several centuries before. He also mentions the birds in the woods as his inspiration. This information confuses the members of the pre-selection committee. A sceptical member declares that he does not expect much of such schooling: '*Oho! Von Finken und Meisen / Lerntet Ihr Meister-Weisen? / Das wird dann wohl auch danach sein*' [Oho! From finches and titmice / you learnt the Master's melodies? / So, your song will be in this vein!][2]. The negative impression is confirmed when the stranger renders part of his song. Hardly anybody finds something catchy in it, neither in the words nor in the melody. They decide to reject the applicant. But one of the masters, Hans Sachs, finds a touch of genius in the singer. He proposes that the knight go over the rules and refashion the song to at least be admitted to the contest. Walther agrees and performs the song, while in the meantime Sachs gives instructions and jots the words down. Later, when Sachs is in the presence of one of the prospective competitors in the song-test – the sceptical and hostile master Beckmesser – he hints at the extraordinary song that has taken shape. Beckmesser manages to seize the text with Sachs feigning ignorance. Beckmesser believes himself to have secured the key to success and performs the song (with his own melody) during the contest. It turns out to be completely discordant and incomprehensible. However, when Walther performs, the same lines and every element seem to have found their natural and logical place, although the result still strikes the audience as unusual.[3] The unusual is recognized as a sign of genius, and it causes no hindrance to winning the test. Purity strikes us initially as a 'false' note.

The story reveals much about German national ideas in the 19th century, and of course those of Richard Wagner, master of '*Zukunftmusik*' (future music), himself. It expresses first the composer's disgust at established rules that tend to strangle genius. Wagner distrusted easy talk about culture and progress: 'Man can only discover his true self, his nature, under layers of cultural sedimentation', he stated in a late echo of Rousseau. The one who succeeds in finding this natural kernel sees through the degeneration of society, discovers his genius and is re-born. This would also imply encountering one's *national* (=natural) identity, through which language and affective bonds to

---

[2] Translation from Peter Branscombe.
[3] Like the real audience in a present-day opera-house because Wagner has masterly succeeded in expressing this feature in his composition of the song's melody.

the environment come into being. Only as part of a nation can a person become a creator.[4] Only then will (s)he start to understand both the sounds of nature and the voice of old masters like Walther von der Vogelweide. Being modern means to become natural and to transcend time and space.[5] Despite Wagner's concerns about the degeneration of his society, his ideas fitted neatly in the German nationalist canon of the 19th century, which also fostered a dislike of rational formulas and glorified intuition. It associated the narrow minds of the Nuremberg masters[6] with other European nations like the English, the French and the Jews, peoples obsessed by volatile money or empty words, whereas the Germans were thought to feed on deeper roots connected with the German Land and the race it had produced.

In contrast with the period of Early Modernity, when the territorial imperative amounted to a system of moral governance and a politico-religious infrastructure, the High-Modern state exposed its citizens to a more uncertain environment in which mobility and personal flexibility counted higher than moral behaviour. Nationalism compensated for the shock of capitalist earthquakes by pointing to something invaluable which was deeply buried in us as members of a nation that had matured in its natural environment. Wagner gave a musical voice to capitalism in his *Ring*, in which industrial sounds of hammering and machines are ascending from the underworld of gnomes, and in which gold causes a lot of trouble. There were several classes in German society among whom modernization raised such ambivalent feelings. The petty bourgeoisie envied the newly rich, who were rocketing to power and standing. The cultured elite continued to worry about their own role as beacon of national progress.[7] This feeling intensified after the 'Wars of Liberation' (against the Napoleonic armies), when local German princes restored 'old regime Europe' while excluding the educated classes in society from politics. The answer of the cultured elite was to escape into the world of the mind or to develop a discourse that was neither cosmopolitan nor rational, like the despised liberalism, but based on a typical German intuitive knowledge. As Liah Greenfeld and others have argued, this form of nation-

---

[4] Joachim Köhler, 1997, *Wagners Hitler. Der Prophet und sein Volstrecker.*, pp. 119-120.
[5] Hans Melderis, 2001, *Raum-Zeit-Mythos. Richard Wagner und die modernen Naturwissenschaften.*
[6] There is some reconciliation with German tradition at the end of the opera where Walther threatens to refuse the Master title and Sachs presses him to acknowledge that the masters ultimately recognised his song as truly German art.
[7] Liah Greenfeld, 1993, *Nationalism. Five Roads to Modernity.*

alism also resorted to the German tradition of *pietism*. Curiously, pietism itself was a spin-off of the Reformation, but also a reaction to the shock of seeing the Christian search for purity of the soul transformed into a bloody search for territorial purity in religious wars.[8] Both influences incited people to rely on an introspective religious world. This lonely turning in towards the self, distrusting social formulas and external religious authorities (the essence of pietism), elicited a need for spiritual communion with those who had gone through the same feelings and revelations, those who were literally *soul*mates. One can understand the elevated meaning of the German concept *Volk* in the 19th and early 20th century from this background. Pietists did not completely abandon the material world; they were active in charity and care of the sick and finally true representatives of the Calvinist work ethic.

In the early 16th century, a handful of German princes had already welcomed the spirit of Protestantism because it seemed to fit in so well with the civic culture of the small state or city-state, which present-day economic geographers would call the spirit of 'growth-coalitions'. Erasmus asking, 'What else is the city than a great monastery?', acknowledged the harmony between work and spiritual life in a territorial community as well.[9] Differently phrased, the statement presented governance and morality as two sides of the same coin. High-Modernity, on the contrary, introduced more sophisticated control strategies that in a way *territorialized* identity. In the first place, systems of general education, set up to answer the new requirements of industrialization, promoted a common language and a communication field that reinforced the state border as a barrier. Second, standardization (like the introduction of a national currency), taxation and infrastructure works made the presence of a central government felt in every corner of the state's territory. This type of 'infrastructural' control[10] enabled governments to influence society in a more indirect way, with financial or laboursaving incentives rather than with coercive measures. Third, loyalty of the citizens had become more important, because individual virtue and coercion were no longer sufficient principles for keeping the social order. State officials had to put more effort into explaining the nature of national regulations or projects like conscription for military service. In a Prussian Edict of 1807, hereditary servitude was

---

[8] K. Hartmann, 1992, *Zwanzig Jahrhunderte Kirchengeschichte. Vom Anfang bis zur Gegenwart*.
[9] John Merriman, 1990, *A History of Modern Europe. Volume I*, p. 104.
[10] Michael Mann, 1984, 'The autonomous power of the state: its origins, mechanisms and results', pp. 185-213.

abolished, and all occupations were declared open to persons of all classes.[11] While neither resolution was implemented very thoroughly, nor did either of them solve the problem of economic inequality and capitalist exploitation, the ruling at least recognized all 'subjects' as 'citizens of a state': people with equal rights. Such messages, matching the nationalist spirit of the times with their aims of general education and commitment to the mission of the state, created confidence in the practices of the state. As Christopher Clark remarks on the Prussian edicts, 'The reforms were above all acts of communication. The propagandistic, exalted tone of the edicts was something new (…) Prussian governments had never spoken to the public in this way before (…) Particularly innovative was [the] engagement of freelance writers and editors as propagandists in the service of the state'.[12]

Whereas the demands made on the citizen in the Early-Modern state – a form of moral composure – were based on tradition, the instruments of control in the new era were rather economic norms, rational argumentation, and ideals of growth rather than 'obedience' as such. Here the territorial shock of becoming a citizen stemmed from the emerging public sphere, a new world of communication that benefited from the increasing movement of persons, goods and information, the extension of literacy to more social classes and the circulation of printed journals. The historian Tim Blanning has made the rise of a public sphere, an idea introduced by the German philosopher and sociologist Jürgen Habermas[13], into the hallmark of political change of the 18th century. This was a major step in the transformation to the High-Modern state.[14] A characteristic literary innovation of this period is the novel. Its appeal lies in the story of ordinary people making decisions – often against traditional morality – that will deeply influence their life, for good or bad. This touched precisely on a new sensitive spot: free decisions could derail one's life, but inaction or living in accordance with handed down morals was not a solution either. One could not escape running risks (Box 4.1). Why might we subsume such dilemmas under the category of *territorial* shock? The reason is that individuals became exposed to a different set of voices and arguments, each claiming authority. National voices did not anymore emulate

---

[11] Christopher Clark, 2006, pp. 327-330.
[12] Ibid., p. 342.
[13] Jürgen Habermas, 1989 (orig. 1962), *The Structural Transformation of the Public Sphere: An Inquiry into a Category of Bourgeois Society*.
[14] T.C.W. Blanning, 2002, *The Culture of Power and the Power of Culture: Old Regime Europe etc.*

> **Box 4.1. Manon Lescaut (1731): an early novel about modern ambiguity**
> The scandalous novel 'Manon Lescaut' by Abbé Prévost is an early example of the enlarged geographical space of human action and the world as a new field of alternatives, or rather seductions. It tells the story of a young man, the Chevalier Des Grieux, who falls in love with a girl (Manon), someone beneath his station. Forgoing the wealth to which he is entitled by his descent, Des Grieux decides to share his life with Manon, but gets into trouble because of her craving for luxury. He even has to condone the swindling of an older and rich man who is lured into a relationship with Manon. Her ability to love people sincerely, including Des Grieux, appears doubtful but is again and again reasoned away because financial disasters are interrupting romantic episodes: 'Do you not see that in the situation in which we are now reduced, fidelity would be worse than madness?' asks Manon.
>
> Deceived admirers, intent on revenge, cause the couple to be imprisoned. The influential father of Des Grieux secures his release but Manon is condemned to deportation to America. Her obstinate lover follows her to Louisiana where, free from the gossip and class distinctions of Europe, they have a short and happy life together. Until gossip catches up with them and they have to flee. On their escape route in the Louisiana wilderness Manon succumbs to fatigue.
>
> The story can be credibly explained as a moral tale and with that the abbot Prévost (who wavered in his own life between religious and secular occupations) came rather well away. But the story also casts some doubt on the easiness of moral counsels and suggests that leaving one's home (existential base) for love is a new challenge that requires special social (and not merely moral) skills.

local authorities, nor did they identify with religion. They appealed to rationality or responsibility, in respect to social change and goals on a wider scale. This very rationality also invalidated religion as a satisfactory territorial marker. Since communication was the hallmark of the new age, language was a logical candidate for shaping people's ties with a territory. But the landscape of language was multi-levelled and fragmented in 18th century Europe. In German courts, even in those where a patriotic spirit ruled as in Frederick the Great's Prussia, French was spoken rather than German. Scholars still spoke Latin. There was no vernacular that smoothly filled the space delineated by political borders. France itself is the telling example of a politically centralized country that nevertheless remained linguistically fragmented until the late 19th century.

The spiritual reaction to this territorial shock was nationalism. Nationalism restored the balance by suggesting that national culture was a genetic feature, something produced not by bureaucratic incantations but rather by a deeper feature in each of us and in our live-space; one only needed 'regeneration' to recover it. Hence, the prospective *Mastersinger* Walther von Stolzing opposed the empty words of a 'bureaucratic' Master with true inspiration nurtured by nature. Anybody who relies on his/her genes (or physique as the early 19th century German gymnastics clubs emphasized) can become superman. This attempt to picture nations as natural facts that root in a territory, or '*Blut und Boden*' (Blood and Soil) as it became known in a less savoury episode in history, is a general feature of nationalist movements, but they do not all assume the same form.

The somewhat exalted shape that 19th century German nationalism assumed had much to do with the inability of the German political structures to honour the new public sphere. In the 17th and 18th centuries, the European economic core moved to the maritime North and West. Martin Swales remarked: '[Eastern and Southern German] towns become enclosed, tied to the relatively stable, institutionalized framework of the guild economy. The skilled master craftsman becomes the elite figure. The "hometown" is not dynamic, expansive, growth conscious. It is inward-looking, distrustful of foreign capital; it is sustained by custom, convention, familiarity, rather than by impersonal laws or economic principles'.[15] This was anything but a hotbed of political and economic change. In Germany the territorial shock was induced by the confrontation with France and its revolutionary events rather than from industry and modern infrastructural power. Kant and Goethe, who enthusiastically embraced the ideas of the Enlightenment, also defused their social meaning by embedding them in an idealistic and individualistic framework. The positive effect that the French revolutionary declaration on the sovereignty of the people might have had on those longing for a more open and democratic society, was further swept away by the reputation of the terror regime in France and its subsequent military invasion into Germany.

The effect of Napoleon's expeditions on German soil was ambiguous. On the one hand, French aggression unified all German classes in the 'wars of liberation'. On the other hand, it gave a final blow to the legitimacy

---

[15] Martin Swales, 1986, 'The problem of nineteenth-century German realism', pp. 73-74.

of the Enlightenment ideal. When the Vienna Congress (1815) finally restored the old European order, the rulers of the German principalities could restore their authoritarian regimes without having to fear an ideologically unified domestic opposition. Since demands for democratic change were routinely rejected, re-territorialization (aimed at a larger German unity) retreated to the minds of painters, poets and composers. It became known as Romanticism.

Box 4.2. Caspar David Friedrich (1823), The Polar Sea: virtual de-territorialization. © bpk | Hamburger Kunsthalle | Elke Walford

An earlier title of this painting was 'Dashed Hope'. A ship called 'Hope', crushed by drifting ice, is just visible. One explanation given for this representation is the mood among German artists who were disappointed with the return to small-state autocracy after the 'Wars of Liberation' ('hope') against Napoleon.[16] However, there is reason to delve somewhat deeper in the spirit of the times to understand Friedrich's artistic quest. Like other works of the painter, this one ignores contemporary 'laws' of representation by omitting a transition zone between foreground and the

---

[16] Jost Hermand, 1982, 'Dashed hopes: on the painting of the wars of liberation'.

> far distance. The spectator is immediately 'flung' into the scene, which causes the feeling of vertigo recorded by contemporaries and often sharply criticized.[17] We may connect this with the feelings of disorientation that this chapter attributes to the new public space of High Modernity.
>
> There is also an explanation from Friedrich's religious quest.[18] The representation of overwhelming natural forces symbolizes the futility of human megalomania and is an attempt toward spiritual communication with infinity. However, even this explanation supposes a (religious) quest that is characteristic of the time and can be linked to a contemporary philosopher, namely Schleiermacher, with whom Caspar Friedrich was acquainted (see further on in this chapter).

In the typical German form of nationalism, the outer homeland had become an inner homeland. Artists tried to transcend time and space in their creative works as a way of escaping the narrow political arrangements of the real world. Richard Wagner and the painter Caspar David Friedrich present fascinating examples of the curious mix of universalistic and particularistic perspectives, which one can find in other nationalisms as well. Its universalism flows from the idea of equality of all people in a cultural and political sense, but its particularism states that one culture is better than all the rest. Yet the first recourse of these artists was spiritual de-territorialization, the creation of an imaginary world in which overpowering natural forces pushed aside the 'authoritarian' framing of the real world, thereby placing 'instinct' above culture (Box 4.2).

The basic nationalist idea of *regeneration,* achieved by delving into deeply buried cultural (declared 'natural') roots, was not a uniquely German phenomenon. In 1758, a young Scotsman, son of a poor farmer, settles in Edinburgh as an instructor working for a higher-class family. He nurses a secret hope of becoming a famous poet; he is talented but knows that his name and undistinguished origin will cross his plans. One day, a relation of the family for which he works asks him to translate a few legends from Gaelic into English. The result is highly praised in Edinburgh literary circles and the young man

---

[17] Johannes Grave, 2001, *Caspar David Friedrich und die Theorie des Erhabenen.*
[18] Werner Busch, 2008, *Caspar David Friedrich: Ästhetik und Religion.* Bush also deduces his religious reading of Friedrich's from the strong geometric composition of these paintings like the application of the 'Golden Section'.

is encouraged to produce more. The result in 1760 is '*Fragments of ancient poetry, collected in the Highlands of Scotland and translated from the gaelic or erse language by James Macpherson*': a collection of songs, allegedly by anonymous bards and preserved by oral tradition. But there is nothing to substantiate Macpherson's claim of their auhenticity. Next, he travels in the Highlands and the Western Isles of Scotland and publishes longer epic poems attributed to a single author like Fingal or Ossian. Particularly, the fake '*Works of Ossian*' (1765), supposedly written by a bard living in the 3rd century before the arrival of Christianity, has a staggering impact not only in Scotland but throughout Europe. These poems seemed to confirm something that intellectuals had already been searching for: the power and authenticity of the local tradition that, in view of the supposed age of the discovery, could be conceived almost as a feature of nature. It proved that culture was not the exclusive domain of the higher classes or the Greek and Romans, but on the contrary something that in its most pure form resides in our own place, in common people. The poems of Ossian unleashed a wave of similar discoveries in other countries, but some of Macpherson's contemporaries already distrusted their origin. As Macpherson aggrievedly notes in the preface of the 1796 edition: 'The eagerness with which these Poems have been received abroad, is a recompense for the coldness with which a few have affected to treat them at home. All the polite nations of Europe have transferred them into their respective languages: and they speak of him, who brought them to light, in terms that might flatter the vanity of one fond of fame'.[19]

    For Goethe and other admirers of these songs it confirmed that – as Wagner would have said – layers of culture veil something of beauty inherent to the most natural human state. This natural state was what nationalists claimed as the origin of a particular group of people (*natio* is derived from *nascere* (L) = being born), but that was all ideology. In reality, at its heart we discover the story of common people, uprooted from a traditional setting, migrating to the cities and encountering other and more influential classes. They discovered that to assume a common culture in all who might cross your path improves your chances in society. Inventing a song already did the magic trick for James Macpherson.

---

[19] James Macpherson, 1765, *The Works of Ossian, the Son of Fingal*, Vol. I, p. iv.

Theories of nationalism

Nationalism is a political doctrine that emphasizes the perennial character of an ethnic group and its entitlement to a distinct territory. Since extolling national virtues and stirring up conflicts with culturally different groups seems to characterize the rite of national identity, nationalism received sustained bad press in the 20th century and was often cast as a lapse into foolishness. When communist Europe disintegrated in 1990, Time Magazine, instead of celebrating the newly acquired freedom of people in Central and Eastern Europe, appeared with a cover showing a map of (Central) Europe disfigured into the shape of a monstrous face. The cover title was 'Old Demon' and in the demon's mouth, the word 'nationalism' was printed. Writing off nationalism as a madness that will pass once people are sober again does not encourage the serious study of nationalist movements. While some solid studies about nationalism already appeared in the first half of the 20th century, an intense academic discussion only emerged during the last three decades of that century. This academic discussion also involved a certain rehabilitation of nationalism linking it to modernity.[20] The course of affairs in post-colonial countries offered a contemporary model of how nationalism could function as a means of integration and stimulus for individuals to dedicate themselves to the development of their country. This interpretation also entailed a more positive view on the role of nationalism in the history of Western nations.

The large array and varied historic conditions of nationalist expressions have made any general theory claiming to explain all cases of nationalism susceptible to criticism.[21] Modernization theory at least has the merit of drawing attention to the role of nationalism as an explanation to a period of structural change in European history and the postcolonial condition elsewhere. A closer look at the manifestations of nationalism at the beginning of the High-Modern period also reveals that nationalism was not in principle based on rousing hate between nations. Initially, and this is something easily overlooked in our time, nationalist expressions reaped *trans*national acclaim. This is illustrated by works of art, such as the poems of Ossian, and also by a revolutionary solidarity among the masses against the *old regime* persisting in different countries. The movement capitalized on a generation gap, as testi-

---

[20] Hans Kohn already emphasized nationalism as a 'liberal' movement. Hans Kohn, 1944, *The Idea of Nationalism: A Study in its Origins and Background*.
[21] Umut Özkirimli, 2000, *Theories of Nationalism. A Critical Introduction*.

fied by the Italian nationalist Mazzini who formed the 'Young Europe' movement in 1834. 'Young' movements sprang everywhere: young Italy, young Egypt, young Turks.[22] The (high) modern *territorial* project went further than stimulating such feelings of being newly born: it implied that *governance* succeeded only by capturing *identity*. This means that new modes of production and new state infrastructural and symbolic powers became intertwined with the identity formation of various groups in society. This is what actually happened in the era usually labelled the Age of Nationalism (1815-1914). It meant that closure receded into the background or, better, that it became a tacitly accepted consequence of governance and identity.

Ernest Gellner, one of the most prominent representatives of the modernization perspective, presented a logical framework for the genesis of nationalism involving three factors: cultural difference within a polity, access to modern education and access to power. When barriers to power or modern education coincide with cultural difference within a same polity, nationalism is the outcome.[23] Gellner also defined nationalism as 'the political principle that holds that the political and national unit should be congruent' or, differently phrased, that political and cultural boundaries should coincide. The apparent message of this statement is that nationalist movements will either: 1. Aim to unite kindred cultural (ethnic) groups that experience division by political boundaries or 2. Cause distinct ethnic groups to attempt secession from a multicultural state. However, for a modernist's taste this inference is stained by 'primordialism', or the assumption that ethnic groups are immutable and distinctive units that persist throughout history. As a modernist would say: this is falling into the myth that nationalism follows the ethnic map. Modernists like Gellner himself emphasized that nationalism *creates* a cultural unit where there was none before, which it subsequently falsely presents as solid tradition. This is not to say that the invented culture has no relation whatsoever to existing traditions. What happened in Western-European nation building, as exemplified by France, was that national unification built on the culture of a dominant core (the language spoken in the Ile de France) while ultimately accepting some regional diversity as part of national identity. Western-European nationalism achieved the assimilation of different cultural groups within a state into a common national culture. Such a thing was only possible by convincing people that acculturation opens an opportunity for

---

[22] Elie Kedourie, 1994 (1960), *Nationalism*.
[23] Ernest Gellner, 1983, *Nations and Nationalism*.

social mobility and access to power. In political systems where this equalizing perspective was absent, for example in the Austrian-Hungarian monarchy, nationalism would eventually lead to disintegration.

In Gellner's theory, the need for modern education (as a function required by the industrial revolution) plays a key role in explaining the rise of nationalism. Education grants access to new jobs in industry as well as services that require flexibility in the workplace and the need to communicate with migrants from different regions. In order to accentuate the new outlook, Gellner contrasts the preceding agrarian-literate society with industrial society. Agrarian-literate (pre-industrial) societies – like medieval Europe – did not feel the urge to eliminate cultural differences. Cultural identity in the sense of literacy or knowledge of Latin was the privilege of a particular class that was supposed to fulfil an essential function in society and that therefore should be kept distinct from other classes. Since solidarity of ordinary people in pre-industrial society naturally went to the family and the village, the greatest fear of rulers was that mobility between classes could make state affairs the plaything of local interests. Administrators and warriors should remain separate from peasants, and the label Gellner chooses for the mechanism that achieves this in pre-industrial societies is *castration*. Castration prevented the eunuchs, the guardians of the Ottoman harem, to interfere with the Sultan's domain by venting to their own sexual interests. Similarly, though not in a literal sense, Gellner chooses that label to indicate a strategy that keeps family interests out of the field of social regulation, like celibacy does in religious communities. Pre-industrial society created classes that had minimal attachment to the masses of common people, by either recruiting state servants from a small class devoted to the state/empire, or by interfering with the process of reproduction as with the foreign 'slaves' in the Ottoman Empire (Chapter 2). Every pre-industrial class monopolizes a specific domain of knowledge and skills – farming, writing, shooting – and there is neither opportunity nor need to switch to another domain in the course of one's life. It is a common misunderstanding, says Gellner, that modern society knows the highest degree of professional specialization. On the contrary, the most advanced specialization occurs in pre-industrial society. Modernization requires the development of general skills (language, mathematics, geography) for people who in the course of their lifetime should have the flexibility to adapt to technological change and should be able to shift to jobs that better suit their talents, even long after finishing formal schooling.

Pre-industrial society is 'addicted to horizontal differentiation', says Gellner, which means segmentation in classes that do not mix.[24] In industrial society the highest possible degree of human exchangeability (equality) is pursued in order to facilitate the 'modern project', based on the idea that anyone can improve his or her position in tune with the general progress of the nation. A modern state demands modular citizens. The most basic way to produce them is general education. People realize that the chance to find work, to develop dignity, self-assuredness and self-respect in society depends on education and that access to education requires a common language and meaning-system, in one word: a culture. People start to develop an interest in culture; 'modern man is not loyal to a monarch or country or religion but to a culture'.[25] Then why this emphasis on the congruence of nation (culture) and state in nationalism? Because only a state is powerful enough to bear and organize the heavy burden of general education.

There has been much criticism of the strong overtone of functionalism in Gellner's approach.[26] Functionalists argue that certain institutions ('organs') or behaviours are required to keep society running. But even if one agrees that social regulation occurs spontaneously, one still observes a missing link if human motivation is completely left out in social analysis. People need reasons in order to act in a certain way, even if such reasons are false. Thus, the question remains 'what motivates people individually to embrace nationalist ideals'? It is not likely that people at the end of the 18th century already grasped the intricate connections between industrialization, general education and culture. Did they really understand that supporting nationalist ideals would improve their position in a changing labour market? Probably not. The connection between a shared culture and personal dignity (success) is apparent, as Macpherson experienced by inventing Ossian's songs, but an awareness of the basics of industrial society would have been beyond contemporary comprehension.

This fits in very well with a second criticism of modernization theory: nationalism also emerged in places and times that were untouched by indus-

---

[24] The term *vertical* differentiation seems more appropriate. Horizontal differentiation in the proper sense of the word – the separation of national territories – is what nationalism produces.
[25] Ibid., p. 53.
[26] Brian O'Leary, 1998, 'Ernest Gellner's diagnoses of nationalism: a critical overview, or, what is living and what is dead in Ernest Gellner's theory of nationalism'.

trialization. David Bell has demonstrated an increasing use of the terms '*patrie*' and 'nation' in speeches and documents since the early 18th century in France.[27] However, at that time these words did not necessarily have the same meaning or scope as in later nationalist discourse. Patriotism could for example mean love for the king, an emotion rather than a political principle.[28] Yet, Bell clings to his interpretation that the increasing frequency in the occurrence of these words indicates an intellectual and material transformation that covers the entire 18th century. Indeed, there were varied and surprising expressions of republican sentiment in early 18th century France. In a 1715 speech on the 'love of country', Lord Chancellor Henri-François D'Aguesseau spoke of 'citizens' and 'senates' instead of subjects and *parlements*. He criticized the monarchy and the newly-deceased Louis XIV and called love of country 'a strange plant in monarchies'. In republics, every citizen would 'from the earliest age … grow accustomed to seeing the fate of the state as his own'.[29] This statement unmistakably prefigured the conception of citizen and state of the nationalist age that started with the French Revolution. Few voices speaking about nation and *patrie* before 1771 (when the *parlement* of Paris was abolished) allow us to infer such a sharp image of the political conceptions of a speaker. It is understandable that critics have sensed a teleological leaning in Bell's insistence on these words as harbingers of a new age.[30] However, if we accept his interpretation, what then is the explanation for this 18th century 'new gaze', the ability to imagine the nation as a sovereign entity detached from the king?

According to Bell, three factors explain the rise of patriotic language in the 18th century: Franco-British rivalry, an extended constitutional crisis and 'new sensibilities associated with the Enlightenment'.[31] The constitutional crisis was rooted in the permanent quarrels between the King and the *parlements*, sovereign courts that were the last vestiges of the nobility's political influence. New sensibilities associated with the Enlightenment stemmed from the changed perspective on the relation between religion and worldly affairs, similar to what happened in the Reformation as answer to the 'territorialization' of authority (Chapter 3). But in France religion and the Church

---

[27] David A. Bell, 2001, *The Cult of the Nation in France: Inventing Nationalism, 1680-1800*.
[28] Ibid., p. 67.
[29] Ibid., p. 51.
[30] See review by Jay M. Smith, in *Journal of Social History* 2003(37)1, pp. 244-247.
[31] David A. Bell, 2001, p. 54.

had been adopted for a long time as tools in state construction for the purpose of legitimizing authority and granting prestige. Religion conferred both an exclusive and a universal character to the French nation. In this politico-cultural configuration, the 'disenchantment' of the world produced by the Enlightenment was seriously destabilizing. The French solution was to endow nation and *patrie* with features and a discourse that were typically analogous to religious or theological talk without relying too much on explicit Christian narratives (such as the Exodus metaphors in England and Holland). An example is the 'cult of great men' introduced in the 18th century. The great men (and some women) from French history offered a mirror for contemplating oneself and an incitement to better oneself. The analogy with saints in the Catholic Church became even more obvious at the institution of the 'French Year', a calendar associating each day with a specific great French character. Another derivation from theological discourse was the emphasis on '*regeneration*' as a potential force in a nation. Looking in dismay at the state of politics in their country, French writers of the 18th century could only cherish hope on a better future by believing in a miracle, something like the resurrection of Christ.[32]

The cultural change in 18th century France was facilitated by material changes. The circulation of French newspapers (which were initially printed in the Netherlands) and the opening of lively Parisian cafés in the 1690s together sparked the creation of the 'bourgeois public sphere', which Blanning highlighted as the crucial catalyst to transformation in the 18th century.[33] All these things preceded the industrial revolution proper but they were in some way or another related to technical change and economic growth. While this material change also occurred in early 18th century England and the Netherlands, where they engendered a similar strengthening of the public sphere and early national consciousness, the French emphasis on 'regeneration' and the dissatisfaction with current society were more in tune with the conception of nationalism as a strongly mobilizing and even revolutionary force. In England, contrastingly, a nationalist or patriotic spirit was rather aroused in confrontations between 'us' (England) and 'them' (the rest of Europe).

In Germany or Prussia, nationalism has also been attributed to an innovative development of a bureaucratic or proto-industrial order. One may

---

[32] Ibid., p. 76, 201.
[33] Jürgen Habermas, 1989 (orig. 1962), *The Structural Transformation of the Public Sphere: An Inquiry into a Category of Bourgeois Society*.

also doubt that frequent use of the factory-metaphor (or cameralism[34]) could have boosted nationalism as a social or intellectual movement. The only thing that apart from education unambiguously contributed to the rise of a moderate bureaucracy was taxation. It was mainly pushed by war, which was a territorially defining activity in the Early-Modern state (closure-identity). War is an expensive affair that originally could be funded by borrowing from private actors or foreign allies. As military operations increased in size, these sources were no longer sufficient and new means of funding, such as direct and indirect taxation (excise), had to be drawn on. The result was a 'fiscal-military state' that showed a greater interest in political stability, economic growth and legitimacy from the public than had earlier systems of governance.[35] It brought a 'revolution in political thinking which prioritized sovereignty with the state as its agent which needed appropriately educated (literate) agents'[36]. This further boosted the rise of the above-mentioned 'public domain', in which state affairs gradually became the subject of open discussion. As Blanning remarks: '[what happened in new public spaces like reading societies, lecture halls, theatres, museums, and concerts, was] the sovereignty of rational argument which prevailed in them, outranking the claims of status or wealth'.[37]

Connecting the fiscal-military state with the growth of a public sphere in only a few sentences undoubtedly appears as a gross oversimplification. There were different national courses in Europe toward a more liberal and open society, but the main point is that this change actually occurred, and secondly that the new 'sovereignty of rational argument' was both liberating and disorienting. Its main result was the individualization of society. Consequently, nationalist arguments filled the void. That this was not merely a matter of 'discovering' ancient songs, but rather something with an impact on the practical routine of the state, can be demonstrated from the German national economists.

The early 19th century writers on the economy, often lumped together as 'economic nationalists', did not extol the virtues of a state but of national

---

[34] See Chapter 3, page 85.
[35] Christopher Storrs Ed., 2009, *The Fiscal-Military State in Eighteenth-Century Europe: Essays in Honour of P.G.M. Dickson*.
[36] Christopher Storrs, 'Introduction: the fiscal-military state in the "long" eighteenth century' in Ibid. pp. 1-22, p. 8.
[37] Blanning 2002, p. 9.

identity.[38] Friedrich List (1789-1846), the spiritual father of the German customs union, noted in his *'The National System of Political Economy'* (1844), 'I would indicate, as the distinguishing characteristic of my system, NATIONALITY. On the nature of *nationality,* as the intermediate interest between those of *individualism* and of *entire humanity,* my whole structure is based'.[39] Economic liberalism, according to List, sees only the exchange value of commodities and ignores the mental and cultural capital accumulated by earlier generations that is included in the products. The Prussian conservative thinker Adam Müller (1779-1829) had already linked the value of commodities with the national identity of the producer: the English industrial product was characteristically stamped with the national hallmark to attest of its high quality, and the German industrial product ought to also acquire a similar national distinction that would render protection through customs tariffs superfluous.[40] The latter unmistakably shows that economic nationalism cannot be equated with the ideals of an autarkical community or extreme protection. Neither is economic liberalism by definition non-nationalistic. Writers propagating economic liberalism often do so with the assumption that this will promote wealth in their own country, which is usually a safe hypothesis in a hegemonic state with a competitive economy. As we know from contemporary world events, a sharp downturn in economic prospects may have a dramatic impact on claims about the economic correctness of a 'universal principle of liberalism'.

The resonance between nationalism and either industrialization or emergence of a public domain is certainly discernable in the 19th century. However, this does not mean that the term nationalism has remained exclusively linked to this period in history. Patriotic sentiments and hate towards foreigners ring out in sources from medieval times all the way to the present and are labelled 'nationalism' according to the view of the author concerned. Yet it was only at the end of the 18th and in the early 19th century that nationalism helped to make the change to a new territoriality. What turned nationalism into a truly *modern* phenomenon is the idea of *mobilizing* society into some special effort that reinterprets internal and external power relations. In-

---

[38] Eric Helleiner, 2002, 'Economic nationalism as a challenge to economic liberalism? Lessons from the 19th century'. Capitals and italics in original text.
[39] Ibid., p. 311. See also Gertjan Dijkink, 2008, 'Nationalism and geopolitics' in Guntram H. Herb and David H. Kaplan, *Nations and Nationalism: A Global Historical Overview*, vol. 2.
[40] Adam Müller, 1808/09, *Vom Geiste der Gemeinschaft.*

ternally, nationalism claimed the equality of *citizens* and their potential capabilities (a born-again experience). Externally, national identity was a competitive factor.

The new infrastructural and symbolic powers of the state

What exactly did the new governance-identity dynamics imply and how did they change the spatial properties of the High-Modern politico-territorial assemblage? Michael Mann has called the type of governance that arose in the 19th century *infrastructural power*. It is part of an oppositional pair: *infrastructural* versus *despotic* power.[41] The modern state does not rule *over* society like a despot but it rules *with or through* civil society. The word 'infrastructure' may be conceived literally: the state, in constructing or planning roads, railways, canals or telegraph lines, stimulates activity and determines its spatial flows, but it does not coerce people in moving along those courses. It is curious that the rise of infrastructural power coincided with a period that is often regarded as the hey-day of the 'minimum interventionist laissez-faire state', the early 19th century.[42] Yet, like today, the state was also necessary for the establishment and operation of a market economy and did many other things that enabled the functioning of an industrial society, such as providing general education, preventing diseases, and providing relief for the poor and crime control.

The very policy of crime control shows how the traditional passive attitude of authorities shifted to a more pro-active mode of operation. It took a while before policing rose up from its passive role in the 'night-watchman state' to the pro-active role of 'beat policing' and investigation. In 1829, the first modern police force appeared in London. 'For the first time the entire city was to be continuously patrolled by men who were assigned to specific territories and whose courses (or beats) were prescribed by their superiors'.[43] But the tracing of criminals was even more sensational. In an article in the *Edinburgh Review* of 1852[44], a reporter tells with amazement how London detectives succeeded in hunting down a burglar by following his track from the

---

[41] Michael Mann, 1984.
[42] Tove Stang Dahl, 1977, 'State intervention and social control in 19th century Europe'.
[43] Jonathan Rubinstein, 1973, *City Police*, p. 10.
[44] Alan Silver, 1967, 'The demand for order in civil society: a review of some themes in the history of urban crime, police and riot'.

location of the offence to his hiding-place somewhere in the East End. The required basic condition for a state doing such activities is not simply a growth in resources (available police officers, weapons or money), but rather the acceptance of the institution of state policies among its citizens. Even more basic than democratic agreement about policies is the state's 'symbolic power' to present its actions as legitimate and beneficial in the public interest. This governance-identity nexus requires some simple invocations about the beneficial aim of infrastructural control: 'We come in peace; these measures do not imply coercion!' Indeed, suspicion about the state's possible abuse of power has inspired resistance against state policies such as the simple counting of the population. Censuses were controversial during the transition to the High-Modern state, and they became so again in the run up to the 'information age'. Such resistance particularly characterizes a stage called 'primitive accumulation of symbolic power' that started in Early-Modernity and turned into 'routine exercise of symbolic power' in the 19th century.[45] Outside Europe, where state formation started later, the primitive accumulation of symbolic power and its concomitant struggles appeared in the 19th or 20th century. Mara Loveman describes how attempts to institute the regular civil registration of births and deaths in Brazil in 1852 elicited violent uprisings that in some rural settlements involved 'hundreds of men' and 'even women armed with knives [threatening] the lives of local authorities who attempted to comply with the law'[46]. Such creation of a birth and death registers aroused two types of alarm: First, it started the rumour that the government would use the register to enslave free people of colour. Slavery continued in Brazil until 1888, but in 1850, only a year before the passing of the Civil Registration Law, the importation of African slaves was banned resulting in new pressures on the domestic slave market. With these events fresh in memory, the evil intentions of the government seemed to be obvious. Second, the strategy of civil registration was suspicious since the church already engaged in recording births and deaths. The new law notoriously prescribed that priests withhold baptism or burial rites when citizens had not complied with the rule of civil registration. This pushed the state into the role of an agent that could frustrate the 'hopes for eternal salvation' of the citizen.[47]

---

[45] Mara Loveman, 2005, 'The modern state and the primitive accumulation of symbolic power'.
[46] Ibid., p. 1667.
[47] Ibid., p. 1671.

The Brazilian events illustrate which qualities a society needs to possess in order to garner symbolic power. First, actions of a state are obviously suspicious when freedom is in such a tenuous position, as in the late slave-holding society. It also explains the exalted tone of the Prussian government after the abolishment of hereditary servitude in 1807 (p. 97 ff.). Second, the collision between the state and the church in Brazil illustrates that a state should sometimes enlist the services of other social agents (like the church) to achieve its aims. Although it seems an admission of weakness, a weak state in this sense has better prospects of developing symbolic power and becoming a legitimate agent than a state that usurps the practices of non-state actors. This is one of the reasons for the non-interventionist approach of the early 19th century European state. Co-opting the practices of local authorities or non-state actors also goes to the heart of infrastructural power itself!

The state and national identity (language, communication) are the major incentives for non-state agents to identify with the state's territory. Apart from transportation-costs, few spatial constraints on human action derive from something other than the state or national identity. Particularly the network of communications (roads, media) that typically breaks off at the state border reinforces the territorial character of all other kinds of social and economic activity in the modern state. Even such transnational organizations as the Catholic Church have adapted their territorial structure to the state and sought state protection wherever available. While industrial production has often aimed for a transnational market, for a long time it remained tied to the labour ethics and rules of engagement of a specific national production milieu. Actually, a firm's playing to the state and other national facilities like banks, trade unions and system of justice may be seen as part of a mode of production that cannot be easily moved across the boundary. The territorial linkage has not become obsolete, even in the age of multinational corporations. There is no simple draining away of all industrial activity to low-wage regions in the world. A 2005 study of the MIT Industrial Performance Center among 500 international companies concluded that companies build themselves up in accordance with practices which correspond to historical differences in the way capitalism developed in such countries as Japan, the United States, Italy or Germany.[48] This sounds like a striking echo of the German economic nationalists from the early 19th century but there is a difference:

---
[48] Suzanne Berger et. al., 2005, *How We Compete. What Companies Are Doing to Make it in Today's Global Economy*.

'The legacies on which companies build are vulnerable, as they get worn down by age and the friction of poorness of fit'. The resources that different capitalist traditions generate get used in new combinations for new objectives: '[National] legacies are not like DNA that goes on indefinitely reproducing familial traits'.[49] While national identity was conceived in the 19th century as something quite static, contemporary thinkers rather tend to emphasize the dynamic way in which a national system of resources responds to global challenges. We have become more accustomed to thinking in terms of complexity and emergent properties of systems, but the conclusion may be the same: a system of resources reproduces itself as a unique territorial identity.

As I noted earlier, one should not think much of the resources of the early 19th century state in terms of money or personnel. The state's strength pivoted on regulation, coordination, and something we rediscovered during the recent financial crisis: sustaining public confidence both in the state (symbolic power) and in the self-regulation of society. Confidence requires transparency of the public sector, and particularly the system of justice. The modern state itself pursued the 'legibility' of society by means of censuses and standardization. An author criticizing 'state-centrism' – a popular target in the 1990s – argued that the simplification of reality caused by the desire for 'legibility' has had destructive consequences. Author James Scott advanced the example of the Prussian state's desire to easily calculate and manipulate wood resources.[50] It mandated the planting of 'standard trees' with a known volume of wood per surface unit given the age of the trees. As Scott argues, this oversimplification of reality ignored the fact that forests had a much wider utility value for the local community, such as supplying medicinal herbs or berries, which now became obsolete. Moreover, the cutting of underbrush and resulting soil depletion eliminated the habitat of birds that normally keep the numbers of harmful insects in check. This is just one example of the self-defeating effect of human intervention on the delicate ecological equilibrium. The harm of such interventions has been widely acknowledged in the second half of the 20th century, but states are not the main actor to blame for it.

Apart from a possible misperception of the Prussian reality (see the remarks about cameralism in the preceding chapter), Scott's criticism of state

---

[49] Ibid., quotes are from p. 280.
[50] James C. Scott, 1998, *Seeing Like a State. How Certain Schemes to Improve the Human Condition Have Failed.*

simplification ignores the fact that legibility was part of an overall desire to make the workings of society (including the state) more visible, not only by means of statistics but also by publicity. Jeremy Bentham, the inventor of the 'panopticon' (1785), wrote in the early 19th century, 'Where there is no publicity there is no justice'.[51] This statement can be applied to both evil machinations of authorities and actions of judges in courts. Newspapers and political pamphlets were instrumental in keeping the public informed about political actions and the ever-ambiguous boundary between illicit and permitted behaviour. It is true that state servants pursuing 'legibility' (in Scott's terminology) of human and natural resources do not like to be in the limelight themselves. It is not surprising that state censorship was a common practice in 19th century (continental) Europe, but the censors themselves were often more sensitive to offensive popular culture (moral issues) than to critical political judgment.[52] The independent political power of autonomous public opinion was not yet recognized or understood.

The interaction between state and civil society in the age of High-Modernity helped to energize the national space by linking governance (economy, justice) and identity (common norms and communicative potency). This, to a certain extent, made closure redundant. In states with overseas colonies (France, Britain) there was obviously no need for territorial conflict within Europe, whereas the German and Italian states looked at identity as a guiding principle of reterritorialization. The (Russian, Ottoman, Austrian) empires followed a different (closure-governance) course that would eventually lead them to collapse or paralysis (an apt description of the Russian Revolution and its Stalinist aftermath). The United States was the only example of a successful (High-)Modern state that in the 19th century preserved the territorial ambiguity of Early-Modern states. Closure is not primarily a matter of territorial shape; it involves international arrangements on spatial movement and influence. Figure 4.1 shows an extraordinary phenomenon: the number of multilateral treaties on cross-border phenomena as war and peace, trade, diplomatic relations and transport caught on (number of treaties begins to rise at the start of the 19th century, but truly soared from the middle of the 19th century). The researchers who collected these data stated that they could not

---

[51] Bentham, 1843, *The Works of Jeremy Bentham*, p. 493.
[52] Robert J. Goldstein (ed.), 2000, *The War for the Public Mind. Censorship in Nineteenth-Century Europe*.

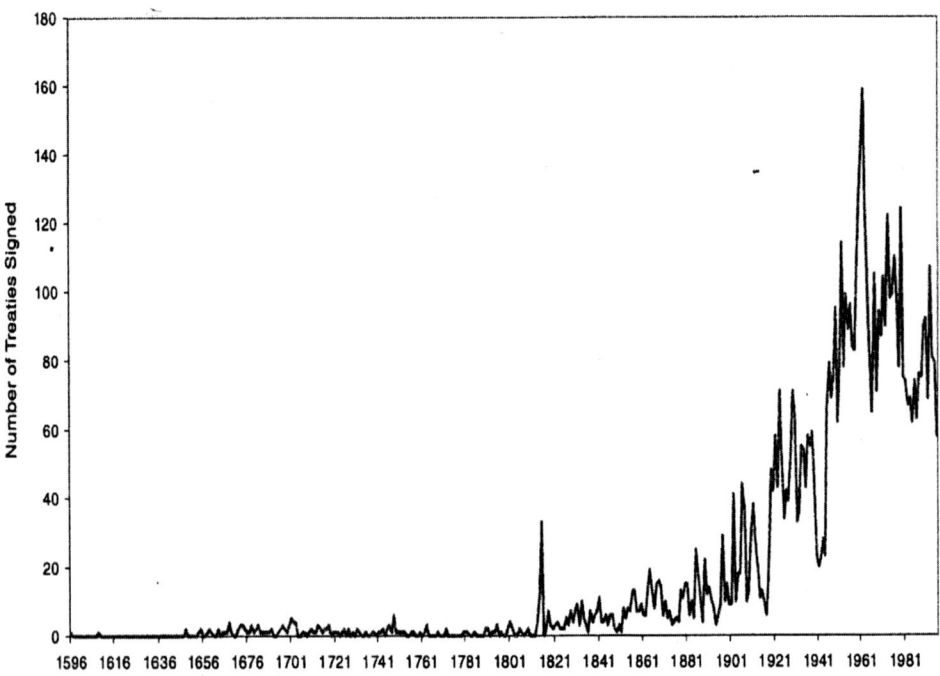

Figure 4.1. Multilateral treaties signed per year 1596-1995. Source: R.E. Denemark & M.J. Hoffmann 2008.

explain the exponential trend from concurrent changes, such as the increasing number of states on the international stage or the expansion of trade.[53] This means that a systemic trend had emerged in which states emphasized mutual connections and developed a shared conception of the nature of boundaries. What had been a historically unique achievement in 1648 at the Westphalian Peace was becoming a 'normalized' response to frictions between states only after 1800. Closure (in the shape of multilateralism) shifted into the passive position that (moral) governance had occupied in the previous period of Early-Modernity. Passive, in the terminology of this book, does not mean immobility but automated response. Diplomacy, which was able to stabilize

---

[53] Robert A. Denemark and Matthew J. Hoffmann, 2008, *Global Diplomacy and World System History: A Network analysis of the Multilateral Treaty System over 400 Years.*

the system without causing conflict, now constituted an essential component of state relations.

Table 4.1 High-modern territoriality

Dimension	Active/passive	(Un)boundedness
Closure Structure of authority. Who are (un)affected by dominant authority?	Basic acceptance of international order (passive)	Multilateral treaties Territorial citizenship
Governance How is a territory / group engaged in turning it into a resource?	Expansion of Infrastructural power (active)	Military conscription General education Public health
Identity How do people attribute *meaning* to power in (territorial) assemblages?	Searching for territorial distinctiveness (active)	Nationalism

A new territorial epoch and its spiritual mark

Few writers were as articulate as Friedrich Liszt and Adam Müller, who developed philosophical statements as a response to the new material and political conditions of the early 19th century. Yet even their views were limited in scope. Overall, it is difficult to find a moral or philosophical message that clearly tackles the broader political and social change of that time. One exception to this is the Italian 'nationalist' Giuseppe Mazzini (1805-1872), a prolific writer who passed a major part of his life in exile. I should immediately add that Mazzini attached a negative value to the word 'nationalism', which he associated with the territorial egoism of states in the wake of the Westphalian Peace, rather than with the civic legitimacy that the leaders of the French Revolution proclaimed. While Mazzini had a distaste for 'nationalism', 'Nationality', on the contrary, was Mazzini's favourite catchword. To him, it meant the new significance of human collectives with a common mission that would together contribute harmoniously to humanity. As Mazzini notes in 1836:

> The epoch that is today coming to an end has sought to achieve the emancipation of the *individual* as its mission: *man* was this epoch's instrument, though this was often poorly understood. The epoch that has just begun, will have the *peoples* as its instrument and humanity as

its goal – that is, its mission – and whoever seeks to move beyond this goal risks committing fateful mistakes.[54]

The distinction between individualism and collectivism (or 'association') became even more central in Mazzini's thought during his exile in 'the capital of utilitarianism, London' (1837-1848). There Mazzini realized that individual 'interests' could never solve the social problems of his time: mere individualism would produce 'charity' at best.[55] In order to pursue *progress*, one needs the nation as an organic whole, held together by a unity of goals and common efforts.[56] The new cry for the 'rights of man', says Mazzini, cannot achieve this goal. Instead, he introduces a somewhat provocative term: 'duty'. It touched a tender spot because, in the early 19th century, 'duty' seemed to chain the working class to the laws of the authoritarian society that Mazzini himself rejected. Therefore, he made a distinction between negative and positive duties. 'Up to now morality has too often been presented to mankind in a negative, rather than an affirmative form. The interpreters of the law have typically said: "Thou shalt not kill; thou shalt not steal". But few, or none, have taught us the active duties of man: how we may be useful to our fellow creatures and further God's design on earth. Yet this is the primary aim of morality, and no individual can reach that aim by consulting his conscience alone'.[57]

It is striking how Mazzini actually articulated the new governance-identity dynamic of the High-Modern Age that was to replace earlier (passive or 'non-territorial') control, which was centred on individual dos and don'ts. Moral government retained an active role in the new territorial epoch, though in a different manner. Whereas in the Early-Modern period prescriptive rules of individual behaviour ranked first, now moral commands related to collective enterprises (education) and the use of national cultural resources dominated. This meant that Mazzini linked human emancipation to the territorial framework of the nation-state, rather than to a cosmopolitan universe of free individuals. At the same time, he declared his allegiance to a transnational

---

[54] From the essay '*Humanité et patrie*' (1836). In: Stefano Recchia and Nadia Urbinati Eds., 2009, *A Cosmopolitanism of Nations: Giuseppe Mazzini's Writings on Democracy, Nation Building, and International Relations*.
[55] Nadia Urbinati, 2008, 'The legacy of Kant: Giuseppe Mazzini's Cosmopolitanism of Nations', p. 18.
[56] 'Nationality and nationalism' (1871). In: Recchia and Urbinati, 2009, p. 65.
[57] 'On the Duties of Man' (1841-1860). In: Rechia and Urbinati. 2009, p. 86.

Law of Humanity that in his view did not need to clash with the pursuance of human progress through distinctive nationalities. For the time being, nationality was the only realistic option for setting people in motion, he suggested, but, apart from this pragmatic argument, nationality was a positive source of strength in terms of emotion, language, and skills. In Mazzini's envisioned world, new nations were complementary to each other and aware of the fact that unilateral violence would ultimately work against their own interests. His *'cosmopolitanism of nations'* was an expression of the multilateral closure type of High-Modern territoriality. The state boundary had become a passive phenomenon, whereas the nation's role and mission now determined the nature of the state in the international system: in short, the new territoriality.

Mazzini's frame of thought represents some of the leading principles of politics in the new age: democracy, the importance of collective goals and equality. Like other revolutionaries of his time, he stuck to these ideals in defiance of the conservative and authoritarian political elite, who attempted to imprison him. Therefore one may conclude that his thought expressed a longing for change rather than a spiritual reaction to an already manifested but confusing territorial change. His thought certainly was spiritual, if only by his frequent invocation of God as creator of the nation.[58] Moreover, his whole approach to nationality was suffused with the spirit of belief and the ritual repetition of words that refer to redemption from a dismal condition: republic, duty, democracy. In one respect, Mazzini's thought was an unmistakable reaction to the cosmopolitanism of individual rights and the materialist individualism accompanying the new (industrial) capitalism.

These monumental changes coincided with an emerging mode of governance that particularly took advantage of individual interests: infrastructural power. The modern state's governance optimally responded to a calculating and growing middle class, but it needed national identity to deflect its potential conflicts and instability. In this sense, nationalism was indeed a reaction and supplement to a changing structure of governance, but this became manifest only in the 19th century.

Not all expressions of 19th century nationalism showed the tolerant cosmopolitanism that Mazzini preached. The phenomenon of national differences that was so positively reviewed by Mazzini and national economists like Friedrich List (p. 113) could easily turn into hate against a foreign 'Other'.

---

[58] Simon Levis Sullam, 2008, 'The Moses of Italian Unity: Mazzini and nationalism as political religion'.

As Liah Greenfeld has suggested, frustrated elites like the French nobility or the class of educated German middle class started the mechanism of 'othering' ('*ressentiment*'). It involved a discursive trick that transformed a feeling of national inferiority towards another nation into a national asset. The strategy was always the same: what one fears or detests the most in one's own society is equated with another nation and what is considered inferior in comparison with the other is revalued as a virtue.[59]

Many commentators, including Gellner, have noted nationalism's shallow philosophical underpinnings, and the connection between this observation and the realist streak in Mazzini's strategy is readily apparent. The central assumption that each nation is different is neither linked to credible facts from the past nor to prescriptions for the future; it is primarily a practical belief. Another spiritual change that occurred in the 19th century sparked profound questions about human existence, while still showing some affinity with the nationalist mindset. The preaching and work of Prussian philosopher and clergyman Friedrich Schleiermacher (1768-1834) was a theological counterpart of romantic nationalism.[60] Although he was not widely known as preacher of a historic religious message, Grenz and Olson describe his influence as '…subtle but pervasive in Western Christianity. He is to Christian theology what Newton is to physics, what Freud is to psychology and what Darwin is to biology. …he initiated a new era in theology… that has lasted at least for nearly two hundred years'.[61] What was this message equal in magnitude to those famous groundbreaking paradigm shifts in science? Schleiermacher's theology has been called 'liberal' because it broke the dogmatic approach to questions about ethics and God. He attempted to connect the principle of human freedom (self-realization) with the idea of man as a social being. The purpose of man's moral quest was to find his/her authentic self, which essentially would entail the spirit of 'humanity'.[62] Encounters with other individuals would be essential in order to define one's personal shape (*Gestalt*) and contribution to humanity. Obviously, this ethics of difference fits the organic concept of society. Similarly, Schleiermacher defined God as

---

[59]. Liah Greenfeld, 1993.
[60] Arlie J. Hoover, 1989, *God, Germany and Britain in the Great War: A Study in Clerical Nationalism*.
[61] Stanley J. Grenz and R.E. Olson, 1992, *20th Century Theology: God & the World in a Transitional Age*, p. 39.
[62] Brent W. Sockness, 2004, 'Schleiermacher and the ethics of authenticity: the *Monologen* of 1800'.

something that should be experienced as a moving force behind reality rather than as an object or person.

Schleiermacher's thinking was averse to religion as something preached in the church. For him religion was a matter of experience (*Anschauung*) and feeling (*Gefühl*). It is not man who masters things but things that take possession of man, and from this confrontation religious feeling is born. There is a clear resonance here with the esthetic method of the painter Caspar David Friedrich, and both knew each other's work. In Friedrich's paintings, direct experience is inescapable; there is no mediating frame or middle ground (Box 4.2). Whether we call this a religious experience or an act of de-territorialization is a matter of indifference to me. The aim is to leave the common ground and to open one's mind to the world as it is: the new public space, however dizzying it may be. While for some, like Richard Wagner, this may count as a return to nature or regeneration, for others it is a religious experience.[63]

There can be no mistake about the new spiritual climate that emerged at the end of the 18th century and that intensified in the 19th century. The emerging philosophies aimed at reconciliation between individual freedom and social reality. In this respect they were a reaction to the shift from authoritarian society to one that pushed the individual into unknown territory, which had to be defined by people themselves. They had to look to each other for answers on how to act, and this electrified the national space as a field of communication. The French state, for example, had been active in forestry projects since the late 18th century, but rural inhabitants resisted those projects, often with little success. Success came in the 19th century when people in the Savoy and Ariège departments mastered the national political vocabulary that had been utilized by rebellious movements in Paris.[64] Today, local community resistance of multinational companies' projects is likewise aided by tapping into **transnational** discourses, for example those produced by the ecological movement.[65] For an average citizen and particularly a rural inhabitant, the territory of the state had not become more real in terms of mobility or tangible connections with places within the state's space. It had become real

---

[63] Both tendencies would merge in Richard Wagner's *Parsifal* (1882).
[64] Tamara Whited, 2000, *Forest and Peasant Politics in Modern France*.
[65] Håvard Haarstad and Arnt Fløysand, 2007, 'Globalization and the power of rescaled narratives: A case of opposition to mining in Tambogrande, Peru'. See also Chapter 7.

as a space of words and images. Benedict Anderson's term 'imagined community' for the modern nation is correct, but we should not misunderstand this term by assuming that the state or the nation has no reality in the practical concerns of citizens.[66]

Territorial shock and the political map

The territoriality of the Early-Modern state pivoted on war and dynastic claims (*closure*), which were often reinforced by religious bonds (*identity*). This system was gradually undermined, partly by its own dynamics and partly by autonomous demographic and economic change. The increasing scale of war, for example, made it difficult to draw on the (unreliable) financial means of the monarch. *Fiscalization* of war was the outcome.

Fiscalization was the closest approximation of 'bureaucracy' in the 18th century state. There were few other forms of state intervention – with exception of the waterworks in Holland – involving a system of routine operations by civil servants. Further growth of state power was only possible when the activities of ordinary people were made applicable to state functioning. In a way, this territorialized the state because it prompted the creation of an infrastructure: education, connections, social data, and health facilities. This shift demanded influence from below (citizens) that might have been a shock for traditional authorities, but it is not the territorial shock as defined in this book.

Territorial shock in the High-Modern age arose from the inability to see the appeal from the state to the individual as meaningful to personal goals or familiar norms. In theoretical terms, it was a conflict between the Early-Modern principle of (moral) governance and the High-Modern implementation of governance (as something linked to active participation of citizens). Even with political representation from below, which for a long time remained very limited and difficult to accomplish, we would still see people being lost in the public domain. Nationalism filled the gap by telling people that there are, in Mazzini's words, 'active duties': that local activities are meaningful in a national space. This, of course, implied a greater public awareness

---

[66] Benedict Anderson, 1985, *Imagined Communities: Reflections on the Origin and Spread of Nationalism*.

of geography and history. Education supplied this information in the shape of fiction and reality in close embrace. For a long time, however, it remained an ideal that appealed little to people in peripheral and rural places.

The resulting High-Modern state was bent on optimizing resources – human and material – within a territory rather than on the mere expansion of territory. This did not however eliminate the desire of states and leaders to take advantage of interstate conflicts to become more powerful or, if possible, to win territory. Now, such actions were inevitably incorporated into a framework of multilateral treaties. The Congress of Vienna (1814), which concluded the period of Napoleonic conquests, produced a map of Europe that remained the territorial clue for the rest of the century. Apart from the disappearance of Poland the most significant changes were the dissolution of the Ottoman Empire in the Balkans and the unification of German and Italian states. Although these unification movements are often labelled nationalist, the impetus actually came from local monarchs or states (Piedmont-Sardinia, Prussia) that were eager to dissolve the Austrian empire to enhance their own power. Others, like the French emperor Napoleon III, supported the Italian independence struggle for the same reason while also acquiring some Italian territory (Savoy in 1860). The nationalist or revolutionary movement in Italy certainly provided an ideological support but not powerful enough to tip the balance. Similarly, Prussian success in the atypical war against France (1870-71) was necessary to accomplish German unification.

It is difficult to maintain that *closure* was passive in light of such massive reorganizations of the European political space as the German and Italian unifications. The great powers in the European state system desired territorial stability, and while identity and language were perhaps not the expected bearers of this result, the desire nevertheless matched the outcome. Identity and language were never sufficient to start a process of political unification, but once established they helped to maintain social and political order by means of an active *governance-identity* compound. In the long run the disaster of two World Wars was not avoided, despite these things. But the outcome of these wars was again a restoration of the established European states (the scale of destruction could easily have motivated a more radical territorial reshuffling). The most radical outcome of the First World War was the further dissolution of the Austrian Empire and the resurfacing of its national minorities in the shape of new Central-European states accompanied by resurrected Poland. Borders were repeatedly redrawn, but this did not affect the continuity of most European states that were present or established in the 19th century.

Chapter 5
CAN THE CENTRE HOLD? TERRITORY IN THE AGE OF LATE MODERNITY

> Turning and turning in the widening gyre
> The falcon cannot hear the falconer;
> Things fall apart; the centre cannot hold;
> Mere anarchy is loosed upon the world,
> The blood-dimmed tide is loosed, and everywhere
> The ceremony of innocence is drowned…
> **William Butler Yeats**, *The Second Coming*, 1919

> From the present time forth, in the post-Columbian age, we shall again have to deal with a closed political system, and none the less it will be one of world-wide scope. Every explosion of social forces, instead of being dissipated in a surrounding circuit of unknown space and barbaric chaos, will be sharply re-echoed from the far side of the globe, and weak elements in the political and economic organism of the world will be shattered in consequence.
> **Halford Mackinder**, *The geographical pivot of history*, 1904[1]

Institutional shift

At the dawn of a blank twentieth century, the British geographer Halford Mackinder produced a shocking image to convince his audience that the world had entered a new phase. After 'the Columbian age' (Mackinder's words), in which European states had freely expanded into the amorphous space of the world outside Europe and no external force ever threatened Europe, a coming world order would again bring medieval conditions. In the Middle Ages Europe had been a self-centred world subjected to blows dealt by a barbaric outer world. Now, in the new twentieth century, such conditions would reappear, because 'every explosion of social forces …will be sharply re-echoed from the far side of the globe'.[2] A hundred years later, when leaders of the most powerful state in the world and officials of the United Nations declared that next to 'evil empires' weak or failed states were the

---
[1] Halford Mackinder, 1904, 'The geographical pivot of history', p. 422.
[2] See quote at the head of chapter.

severest threat to world order, Mackinder's vision seems to be curiously realised. One may also assert that not much has changed since his doleful prophecy; his statement merely shows that 'globalization' was already being experienced at the beginning of the 20th century, and according to many an author, even much earlier.

Globalization, understood as increasing interrelatedness of different parts of the world or the supply of products from global sources, is not a useful concept to define a new era that is supposed to have started in our own time (in 1980). The gradual increase in global exchange of products and global distribution of people spans millennia. If any quantitative approach produces ambiguities, we may at least establish a qualitative geopolitical change in the current period: the forces that influence the global (economic) system no longer exclusively originate from the West. Now, such forces emanate from Asia and other continents alongside the West. Some authors use the inappropriate[3] term *reversed globalization* because countries that until recently were at the 'receiving end' of the globalization process are now active contributors. This may be interpreted as Mackinder's vision coming true, although the distinguished British geographer was rather thinking about security risks ('explosion of social forces') for Europe, and particularly the British Empire. Obviously, social forces outside Europe or the West now have an autonomous impetus with a worldwide reach. Even strong states are vulnerable to failing states: a notion an imaginary visitor from Mackinder's 'Columbian age' (1500-1900) would have trouble understanding. Yet, a new vulnerability is precisely one of the distinctive marks of what we name – for want of a better term – the 'era of globalization'. Its shocking characteristics include the migration of production to economically weaker countries, the migration of the global down-and-outs to the global 'core', the targeting of busy areas by terrorists, information wars with hidden enemies and the contagious effect of corruption and disease that were previously considered as characteristic of backward areas.

The new discourse on security confirms that threats have become less tangible over the course of time. They have become distinctively transnational (in the sense that they are both external and internal). They are materially light but at the same time produce massive effects. Once revealed, they cannot be controlled because they continuously appear in a new shape. The

---

[3] Inappropriate because 'reversed globalization' literally means increasing closure and isolation of countries or regions.

Italian sociologist Alberto Melucci introduced the term 'symbolic multiplier' already in the 1990s to express the fact that power in the contemporary world no longer depends on commanding a huge administrative apparatus or army, but rather on choosing the right symbolic tools that have the power to manipulate the masses.[4] Terrorists, to be sure, use military tools like guns and explosives, but their aim is to immobilize or humiliate the enemy psychologically rather than materially. The most strategic weapons of the Islamic State were videos and a 'glossy' web-magazine (*Dabiq*). Other thinkers, like the German sociologist Ulrich Beck, have preached the coming of the 'risk society' in which neither social status nor power or wealth could guarantee immunity against the risks inherent to a human-created world that has no precedent.[5] His nomenclature choice for the new era is 'reflexive modernity', which means that, for the first time in history, we have to think about the destiny of humanity. We cannot banish dangers and choose to imperil others while saving ourselves with impunity.

War statistics show in a different way that threats are coming from within rather than from without. Since 1945 the number of wars between states decreased, whereas the number of intra-state (or civil) wars increased. This does not exclude the possibility that external actors, either outside states or private groups, are involved in such conflicts ('proxy wars'), but they operate in a way that differs from states on the warpath in an earlier era. At the same time, the sovereign state has never been more successful than in the last half-century, witnessed by the many new states that have appeared on the world stage. It should not be taken for granted, however, that the new states simply reproduce the strategies on territorial dimensions (governance, identity, closure) of longer-established states. New states may be frail, and when faced with challenges such as volatile markets or uncertain internal support they may adopt modes for coping deemed uncivil in parts of the world with a longer state tradition. However, the actions of executive government branches in these long-established states are drifting in the same direction under the influence of new threats. Democracy became a tremendous ideological weapon in international relations, but it simultaneously lost its essence in a world that has virtualized politics.[6] The manifestations of this precarious situation can be observed in the modern world: politics have become a game

---

[4] Alberto Melucci, 1996, Challenging Codes: Collective Action in the Information Age.
[5] Ulrich Beck, 1986, *Risikogesellschaft. Auf dem Weg in eine andere Moderne*.
[6] Andrew Wilson, 2005, *Virtual Politics. Faking Democracy in the Post-Soviet World*.

in which parties are invented to confuse the electorate, election results are manipulated, and playing on public opinion is more important than a well-considered programme or social analysis. For ordinary people, the difference between words and reality has become more elusive. Andrew Wilson coined the phrase 'virtual politics' in 2005 as a description of political practices in the post-Soviet world. A decade later, the phrase got a fresh implementation in the electoral struggle and presidential practice of Donald Trump, a president who, with impunity, refuses to release his tax returns and has not appropriately separated personal business interests from his public office.

In addition to the unprecedented role of risks, the new age is also marked by institutional changes. States face three types of institutional challenge in the Age of Globalization: First, as explained above, private agents have become more powerful. In 2011 only 20 countries in the world were richer than Bill Gates, founder of Microsoft. Now, there are many more people at or close to that rank. Second, non-state agents easily move their base of operation across state boundaries. Multinational firms are perhaps not as footloose as is often assumed[7], but social movements have strongly benefitted from the rise of Internet. Third, state and non-state actors have increasingly powerful means at their disposal to avoid juridical institutions and practices. *Wikileaks's* publication of secret diplomatic documents containing information about unsavoury and unlawful actions showcases this point. Further, the revelations in the *Panama Papers* about how politicians and firms funnel money abroad to 'safe havens' is another. The adaptations that states are making to these new conditions are as follows.

First, states have delegated their duties to private agencies or local governments that are supposed to engage in forms of governance that rely on public-private cooperation. This approach has created authorities or firms that in certain aspects behave as independent states, including engagement in a kind of foreign politics ('para-diplomacy'). This violates one of the basic principles of the 'Westphalian state': the joining of all governmental contacts with the outside world through one central channel: the central government or the foreign office.

Second, states have set up international organizations in order to control undesirable transboundary movement. These vary from supranational organizations like the European Union to common arrangements for trade reg-

---

[7] See the remarks on the new infrastructural and symbolic powers of the state in Chapter 4, pp. 111-118.

ulation and arbitration (WTO) or patents. We use the generic term *supranationalism* for such arrangements even though they are not always equal the advanced organizational shape of a European Union.

Third, states appropriate non-juridical forms of control and regulation and evade accountability by involving (passively or actively) transnational private agents or parts of state-agencies of other states. This may include the operation of secret prisons or setting up of forms of intelligence gathering that violate human rights. Such strategies necessitate a power concentration in the executive, which at first sight seems to contradict the idea of a power shift to the private sector. But the executive itself is acting like a private corporation in a cooperative framework with other agencies. When the executive uses external agents to increase territorial coherence and security I propose the term *transnational regime*. In the age of symbolic multipliers this does not necessarily require the capabilities of a 'superpower'.

Table 5.1 Explosion of governance (IGO=International Governmental Organization; INGO=International Non-Governmental Organization)[8]

	STATE		PRIVATE	
	National	**Transnational**	**National**	**Transnational**
LEGIBLE	A. Government Central state Local government	B. IGO Supra-national gov. Security organization Trade arbitration	C. Civil society Heritage foundation Philanthropy	D. INGO Social Movement Multinational firm
NON-LEGIBLE	E. National regime Shadow state Security state	F. Transnat. Regime Secret prisons Flexible sovereignty	G. Nat. network Shadow economy Rebel militias	H. Transn. Netw. Private govern. Terrorist network

Table 5.1 summarizes the institutional field in which states operate (classic model defined in cell A). It is a three-dimensional progression in which one moves from state to private, national to transnational and legible to non-legible. The label *legal* (vs. *non-legal*) would also apply to most horizontal differences, but not all forms of organization in the second row are illegal: the shadow economy involves activities like barter that are legal, but not recorded. Therefore, the term legible is preferred. From the classic state (cell A),

---

[8] Some elements of the model are inspired by a table of Latham, Kassimir and Callaghan. Robert Latham, Ronald Kassimir and Thomas Callaghan, 2001, 'Introduction: transboundary formations, interventions, order, and authority'.

one moves (in 3D) to the right in the direction of private (non-state) and international institutions, and downward to non-legible institutions. Consequently, A's antipode (diagonal) is H ('Transnational networks'). This cell implies neither effective state nor controllable activities nor territorially bounded institutions. This applies to international crime syndicates and terrorist networks (Al Qaeda), but also to non-criminal activity like systems of arbitration, which are often erected by international firms (parties agree to acquiesce to the judgment of self-appointed authorities, in other words, private governance). State agencies will feel more at ease with the institutional phenomena in the 'legible' row of table 5.1 since the state is (indirectly) responsible for drawing up laws and directly responsible for maintaining juridical practice. In this sense the non-state legible sphere (C, D) is liable to democratic control as well. However, democratic control in a populist or demagogic vein may also lead to (illiberal) action against transnational movements (D), as actions against international social movements in Russia and China as well as actions against the Central European University in Hungary have shown. State power may also escape democratic control when members of the executive indulge in secret activities. A 'shadow state' (cell E) develops when state officials use their status merely to pursue personal gain. It may ultimately result in the erosion of all public service while the formal structure of the state is simultaneously and meticulously upheld (because state authority makes or breaks personal fortunes of the statesmen). The 'security state' (E) is imposed as a response to an emergency, but since 'crisis' assessment is liable to political manipulation, the borderline with corruption may be thin. Unbounded power for the executive is obviously practical when he or she (and his or her government) is pursuing personal goals. The most recent example of a shift to the non-juridical sphere in a Western state is the suspension of rights in the context of the war on terrorism (torture, extraordinary rendition). These measures represent a less comprehensive break with democratic norms than the security state which is activated in response to a massive attack or civil war. The war on terrorism may involve only national actors (E) or a transnational regime (F) with secret operations, as with the network of secret prisons that the American CIA ran in several countries.

The list of institutional forms in table 5.1, particularly the non-state category, does not imply that any one of them is a unique product of the late-modern period. INGOs (D) for example have a long history (in the shape of religious orders or the Red Cross (1863)). A historic analysis by Thomas Davies has shown that there is no linear increase in political activism of INGOs over the course of time, although their numbers have certainly risen in the 20th century.[9] There were cyclical surges in INGO activism during the turn of the 20th century, the 1930s and the 1990s. Geopolitical changes such as the end of the Cold War appear to be obvious causes, but there are many other factors involved, such as new technologies. The rise of information technology, which is characteristic of the current era of globalization, may produce a new INGO surge. However, no institution in table 5.1 is a brand-new product of the current era. It is the variety and the versatility of these institutions that is characteristic of our age.

Ideal-type A (table 5.1) is that of perfect closure, a container in which no control is transferred or lost to institutions outside the container. Peter Taylor has corrected this assumption by remarking that the container model is (onto)logically inconsistent in view of the requirement that the existence of territories relies on a broader accepted principle of division and external recognition: inter-territoriality or inter-stateness.[10] But, apart from such ontological considerations, state-containers are actually 'leaking' because their boundaries do not coincide with the mechanisms that produce contemporary wealth and security. The new institutions that characterize these inter-state wealth containers (trade blocs) or security containers (military alliances) can be equated with cell B. However, a state's dependency on extra-territorial resources is not completely remedied by a quasi-territorial shell of inter-governmental arrangements. States (or their central executives) also enlist the help of foreign organizations (either private or foreign state agencies) that work outside the authority of juridical or democratic institutions (E and F) because this is seen as most effective.

The emergence of a non-legible sphere does not necessarily mean that the 'centre cannot hold' (if we equate centre with state), only that states are less accountable and more easily ignore negative effects for specific groups of citizens and for other states. Equality is the first victim of globalization, and international stability the next. In this way W.B. Yeats' remark, in a poem that links the terrible experience of the Great War with the end of a religious era (opening of chapter), that 'mere anarchy is loosed upon the world' may apply once more. To make this more concrete the following pages provide a

---

[9] Thomas Davies , 2013, *NGOs: A New History of Transnational Civil Society*.
[10] Peter J. Taylor, 1995, 'Beyond containers: internationality, interstateness and interterritoriality'.

number of examples from the second row of table 5.1 (E,F,G,H). We will begin with the phenomenon that is a favourite villain in populist crusades against globalization: supranational authority (B) and the ambiguous role of non-governmental organizations (C,D).

Supranational governance (B)

If there is any change in governance that has raised unease in the past decades, it is supranational government. It arrived most tangibly in the shape of the United Nations and the European Union. Unsurprisingly, it evoked both hope and hatred, not to say extreme paranoia among some actors. Supranational governance literally means that decisions can be made 'independently' by a higher institution regarding large-scale borders crossing questions. Such decisions do not always coincide with specific national preferences. Although a majority vote is required to settle radical global (UN) or regional (EU) policies and certain policy changes are subjected to veto-rights, the practical outcome may turn out to be difficult to digest by certain countries and cannot be ad hoc discarded by the afflicted country. In such cases, resentment against a supranational body may arise, but the need for and benefits of supranational governance on other issues usually overrides fear of losing national sovereignty. After all, states, unlike subnational governments vis-à-vis the state, are free to exit a supranational organization.

The trauma of a 'world' war, the fear of nuclear proliferation, decolonization and political instability in the 'Third World', human rights violations and more recently the global climate crisis are all problems that demand international coordination and the establishment of offices that gather global information and perform interventions on a daily basis. The impact on national sovereignty is more tangible for EU member countries than for UN members, as shown by the EU's open (internal) borders, common currency, guarding of the external frontier, common laws (the '*acquis communautaire*'), Court of Justice, a High Representative for Foreign and Security Policy and a European Parliament, etc. This has elicited the comparison of the EU with a (federal) state or an empire. The comparison between the EU and a federal state falls short for three reasons: There are no unified federal enforcers (police or military, with the recent exception of FRONTEX), there is no individual federal taxation and there is no election of the federal executive. If we omit the historical example of the European colonial empires, the term empire makes some sense: internally because of the presence of autonomous governments (provinces) with their own 'rulers', and externally because of the EU's interference in bordering regions. Despite these similarities with empire, there is no strong European core with civilizational and power dominance over the rest (which reflects a history of conquest).

The most sensible comparison to the EU with a political body known to us is not the medieval European idea of empire, which had not yet incorporated the idea of territorial sovereignty, but the Holy Roman Empire as it survived the great landmark of territorial sovereignty, the Peace of Westphalia.[11] The Holy Roman Empire of the German Nation (HRE) was a political umbrella over the multitude of German principalities and larger German states that existed for almost a millennium, including (part of) the modern era. The Empire's princes, cities and electors[12] recognized the Austrian Emperor as their monarch and participated in a form of governance on different levels. There were empire-wide institutions like the Imperial Diet (*Reichstag*), which had wide-ranging legislative powers regarding internal order, peace, security and economic regulation. The implementation of these decisions was delegated to Imperial Circles (*Reichskreise*), administrative districts (not necessarily territorially contiguous), led by one to three important princes and checked by assemblies. The Empire also had a judicial structure with two imperial courts dealing with territorial conflicts, conflicts between rulers and their estates and social and religious conflicts. The more powerful territorial princes obviously resisted the local administrative arm of the Empire, but attempts to reinforce the principle of territorial sovereignty in their state – by moving themselves to the position as the representative of their state – only strengthened the *Reichstag* and the Emperor's influence. The need for counterbalancing influences may explain the relative stability of the HRE, which persisted until the beginning of the 19th century. Stability was further shored up by the external powers who were party to the Treaty of Westphalia, of which two, France and Sweden, had direct interest in preventing the rise of a strong centralized power in the heart of Europe. Another stability factor was the history of religious conflict in this area that induced willingness for com-

---

[11] Ronald Axtmann, 2003, 'State formation and supranationalism in Europe: The Case of the Holy Roman Empire of the German Nation'.
[12] Electors were the princes from dynasties and archbishops who traditionally 'elected' the Emperor such as the Duke/Elector of Saxony, the Margrave/Elector of Brandenburg or the Archbishop of Trier. The Holy Roman Empire had seven to nine electors.

promise and mediation reminiscent of consociational politics[13] in modern states. In the end, the French military campaign in Germany ended the HRE in 1806. After the fall of Napoleon, strong German states like Austria, Prussia and Bavaria were bent on strengthening rather than sharing their power.

While most states have historically been unified by means of (gentle) force and only later attempted national legitimation, supranational governance has been achieved essentially from below by states with an established national identity. Exit is therefore a realistic option, but no easy decision. Leaving the EU would undoubtedly incur costs and damages of which the obscurity itself is a substantial deterrent. Both EU membership and leaving cannot be adequately explained as a calculation of costs and benefits, but rather follows what March and Olsen have called a 'logic of appropriateness': the pursuit of a purpose more associated with identities than with interests and with the selection of rules more than with individual rational expectations.[14]

> **Box 5.1. Brexit and the logic of appropriateness**
>
> On June 23, 2016 a small majority of British citizens voted in a referendum for the option of a British exit from the EU. There has been much discussion since about the question if British voters realized what they were doing, but the magic word of 'sovereignty' offered beyond any doubt the required 'logic of appropriateness'. As a farmer woman told in a radio interview: 'We will be finally delivered from that meddling body in Brussels'. But, asked the reporter, 'You will lose the generous EU support for agriculture?'. 'Yes, but that makes no odds against our independence, and we will be compensated in some way'. The idea that local institutions show a greater solidarity with your interests, however, is not warranted. Similar disgust in Britain based on the idea of sovereignty has been targeted at institutions like the European Court of Justice and the European Court of Human Rights (not an EU institution). Ironically, these

---

[13] Consociationalism is a correction on the representative democracy with the aim to give different (cultural) groups in a society an equal say in important state decisions. It may be defined as a non-territorial variant of federalism. Consociational democracy (Austria, Netherlands, Belgium. etc.) offers representatives of such groups a strong voice in policymaking but leaves the final decision to a representatively chosen Parliament. Arend Lijphart, 1969, 'Consociational Democracy'. Jan Markusse, 1997, Power sharing and Consociational democracy in South-Tyrol.

[14] James G. March and Johan P. Olsen, 1998, 'The institutional dynamics of international political orders'. p. 951.

> institutions were much busier with judging other European nations with values that originated in Britain than with actual interventions in the British judicial process.[15]
>
> Other powerful motives for Brexit originate with the fear of competition in the labour market by immigrants and ineradicable xenophobia. There is a deeper sentiment, however, that associates the entire supranational enterprise with contempt for it from the classes that feel themselves powerless vis-a-vis the new mobility and information society.[16] It is the sentiment of the arsonist that wants to destroy something of value, not because (s)he hates it but because the opponent cherishes it.

Supranational institutions are clearly an easy target for public anger that actually springs from other developments, such as the changing structure of employment, growing inequality of incomes, mobility of multi-national enterprise and the political turbulence in the Middle East and Africa.

Private legible organizations (C,D)

As mentioned above, the actions of Non-Governmental Organizations may not square with the aims of a national government, but they are nevertheless controllable. Only a transnational organization (INGO) like Amnesty International or Greenpeace may continue its actions if it is thwarted or forbidden in a specific state. In the age of the Internet this means that INGOs, even after being banished from one state, may remain powerful activists capable of mobilizing international opinion. NGOs, on the contrary, are more likely to act as a support in democratic governance rather than an impediment. This role is often discussed in terms of the significance of 'civil society' for a healthy democratic system. In line with discussion about the symbolic powers of the state, explained in connection with the Brazilian census from 1852 (page 115), the enlisting of non-state organizations may reinforce the legitimacy of a state (in contrast with policies that send a message of strength by usurping the practices of non-state organizations). Here the vicissitudes of a contemporary 'strong' state, China, may be illuminating.

---

[15] Jon Henly, 'Why is the European court of human rights hated by the UK right?'. *The Guardian* 22/12/2013.
[16] Zadie Smith, 2016, 'Fences: a Brexit diary'.

After two decades of all-consuming one-party rule under Mao Zedong, a period in which the state strangled any opinion that deviated from the party line (culminating in the notorious Cultural Revolution), a period of reform dawned in 1978 under Deng Xiaoping. It brought greater freedom of expression and organization, including in the economic sphere. Although the autocratic system was not dissolved, the new regime's economic success was astonishing and even moved many Western observers to accept the idea that autocratic government and economic success were not so antithetic as previously believed. Meanwhile, increasing material satisfaction together with a greater tolerance from the state apparatus stimulated a pursuit for spiritual fulfilment among the Chinese citizens. This opportune environment fostered a mushrooming of religious movements which were tolerated as long as they did not undermine the authority of the communist party. However, problems arose along with the religious fervour. For example, when Catholic religious groups acknowledged the authority of the pope, there was a need to defuse the situation by adopting a non-threatening Chinese version of Catholicism. This method was common and was replicated to deal with other 'transnational' movements as well. Indeed, the communist Party's fear of interference with their authority was not limited to spiritual movements of foreign origin, as the story of the Falun Gong shows.

The Falun Gong movement developed on the basis of a traditional Chinese philosophy about harmonizing bodily functions and meditation (Qi gong). Initially, the claims of increasing personal health and even mental capacities did not alarm the authorities; on the contrary, there was much appreciation of its positive effects against the background of inadequate health services and the movement's truly Chinese roots. But its success, thanks to the energetic leadership of Li Hongzu, was ultimately a factor in its elimination by the authorities. The sectarian character of the movement, with its claims of enabling superior insight, elicited heated discussions with opponents and accusations of deceit. Such occasions gave authorities the opportunity to make some arrests, which subsequently provoked a Falun Gong sit-in demonstration directly in front of the government buildings (1999). It was a completely peaceful protest, but at the same time such a display of will power against the government that the latter lost its balance and started to persecute anyone affiliated with the movement. These measures did not really eliminate the movement; it continued underground and abroad where it became a success story within the Chinese diaspora.

The Falun Gong is one of the examples Carl Minzner adduced to support his thesis that China is on its way to a severe political collapse because it has not developed the institutional flexibility to absorb the forthcoming economic and demographic crises.[17] Let us here restrict this wider issue to the consequences of crudely suppressing a national movement like Falun Gong. Minzner specifies three harmful consequences: driving the movement underground at home, its radicalization outside China and its transformation (abroad) in a variety of institutional manifestations. Going underground made Falun Gong 'illegible' but not inactive. It kept appearing by hacking into state cable channels or by disseminating hidden messages in public places. The radicalization abroad engendered a transformation from action against the specific political leaders that were directly responsible for the persecution of Falun Gong members in 1999 to a much broader attack on the Chinese political system as an anti-civilizational force. The institutional extension transformed the diaspora, a 'group of elderly retirees gathering in parks', into an entire news machine that fed newspapers and websites (at least in the English-speaking world) with scandals like party corruption, state-sponsored organ harvesting from prisoners and internal party struggles. This demonstrates that autocratic states struggle to cope with movements in the information age; a 'weak' approach would probably reap more success.

Transnational regimes (F)

Flexible sovereignty

Flexible sovereignty is at stake when states feel obliged to control people and material conditions situated outside the boundaries of the acting state proper. So-called 'diaspora engagement' targets former state nationals who have emigrated to another country. While this may violate the sovereignty of another state, it is often tolerated when the issue touches on matters of security or mutual interest. However, the nature of the action usually requires that it remain outside public scrutiny. Another, perhaps more benign, form of external state interference is the attempt to mobilize a diaspora in order to secure either transfer of know-how and remittances or political lobbying in favour of the country of origin.[18] The line separating mutual interest and undesirable

---

[17] Carl Minzner, 2018, *End of an Era: How China's Authoritarian Revival is Undermining its Rise*.
[18] Alan Gamlen, 2006, 'Diaspora engagement policies: what are they, and what kinds of states use them?'.

foreign interference is of course very thin as exemplified by the political pressure of the Erdogan regime on Turkish inhabitants of Germany and The Netherlands.[19]

Already in 1955 the political scientist Morgenthau recognized the 'dual state' nature in which politics was conducted in the United States.[20] One side is the democratic state that acts according to the law and is open to public discussion and influence; the other side is a 'security state' that monitors and controls the former. The 'security state' acts in matters of life and death. It is always present and will act in case of 'emergency'.[21] After September 11 much of public life in the US was securitized. The formulation of the infamous Patriot Act was crammed into only six weeks, after which it was accepted with overwhelming support in Congress on 25 October 2001, even though it largely contradicted the 'sacred' Fourth Amendment in the American Constitution: the right of people to be secure against unreasonable searches and seizures conducted by the government. Hundreds of people were detained without being charged or tried and without the government giving relatives any information about their whereabouts.

In the wake of September 11, securitization has spread unchecked in the Western world. It was not a mere rational reaction to a concrete threat, such as a visible advancing army, but instead involved a good share of the 'evangelism of fear' preached by the United States. On instigation of the US, other Western states and members of the NATO alliance started to adopt measures that fell more in the model of the security state than in that of a democratic one. In the Netherlands suspects could not be pre-emptively prosecuted, even if they were deemed to be members of a radical Islamic network (the so-called 'Hofstad group') and possessed damning documents containing topographic descriptions of public buildings or video testaments in the style of the Middle-East suicide bombers. In 2006 the law was changed to the extent that any trace of a discussion or design of a terrorist attack, whatever its imaginary status, was considered a penal offence. With such

---

[19] Gertjan Dijkink and Inge van der Welle, 2009, 'Diaspora and sovereignty: three cases of public alarm in The Netherlands'.
[20] Hans J. Morgenthau, 1962, *Politics in the Twentieth Century. Vol 1 The Decline of Democratic Politics.* (Chapter 29: 'The Corruption of Patriotism'), p. 400.
[21] Ola Tunander, 2004, 'Securitization, Dual State and US-European geopolitical divide or: the use of terrorism to construct World Order'.

measures, even the designer or player of a violent video game situated in the real world might become indicted with terrorism.

Another disturbing implication of the security state is that it easily facilitates the transfer of personal information on citizens between states in such a way that there is no public control over the nature and quantity of information exchanged. While states often tacitly agree to the transfer of information, there are public spheres that may compel states to retrace their steps. Whatever the occasion for the criticism of the democratic shortcomings of the European Union, European Parliament members often step in to denounce practices that infringe upon individual rights. One example is the Brussels-based international money transfer company Swift's secret transfer of private information (see Box 5.2). In actuality, the EU can compel Belgium to stop such practices because they violate EU agreements.

> **Box 5.2. EU concern at US data transfer** (BBC 31-1-2007) [22]
>
> Last year, it emerged that a private company, Swift, which handles up to 11 million money transfers a year, had been passing information to the US authorities in violation of EU privacy rules - a finding Swift disputes. Now MEPs want to know whether that data was fed into the US Automated Targeting System (ATS), which profiles possible terrorism suspects.
>
> The US Department of Homeland Security (DHS) released information about the targeting system in December (2006), explaining that it was intended to detect high-risk individuals previously unknown to the law-enforcement community. It gives anyone entering the US a numeric score, a measure of the risk he or she is thought to present, which the DHS said could be shared with state and local police and foreign governments.
>
> 'The European Parliament is fully supportive of proper co-operation across the Atlantic in fighting terrorism... What we don't accept is that there should be misuse of data'. Liberal MEP Baroness Sarah Ludford told BBC World Service radio's World Today programme. She said that ATS went 'way beyond anything we had been led to understand the information would be used for'.

A similar vigilance was exhibited by the Council of Europe, which in June 2006 published a report on the so-called 'extraordinary rendition flights'. The

---

[22] http://news.bbc.co.uk/2/hi/europe/6315893.stm (last accessed March 4, 2018).

report suggests that European governments have colluded with the CIA over the transport of terror suspects to (possibly) secret jails in Europe as well as to torture sites in such countries as Syria, Egypt, and Afghanistan. Investigators of the European Parliament have registered more than 1000 undeclared flights.

Rendition also revealed another way in which states made closure flexible: the way they treat persons with dual citizenship, particularly people of Arab descent. A notorious case was that of a Syrian born Canadian citizen, Maher Arar, who was arrested in New York in 2002 and subsequently flown to Jordan where he was extradited to Syria. Arar was tortured for over 10 months in a Syrian prison and finally released in 2003. The uncomfortable truth is that Canadian intelligence agencies seem to have collaborated with their Syrian counterparts in order to get specific information from the prisoner. This is in clear contradiction with the tradition of citizenship, which requires the protection of nationals abroad through all diplomatic means.[23] Apparently double citizenship has created some room for a flexible interpretation of a government's responsibility.

While states (like Canada) have shown remorse about such cases, the ultimate conclusion is that ominous world events may induce states to shift traditional norms of sovereignty and citizenship. However, this continuous realignment of the relations between states is not a sign of the disappearance of states. As Stasiulis and Ross conclude, 'While characteristically diffuse and often lacking territorial concentration, flexible sovereignty does not follow from, or lead to, weakness or emasculation of state power'[24]. This judgment, of course, does not consider democratic accountability or social unity as factors strengthening state power. It equates the state with a strong executive that may at worst be an authoritarian regime or a dictator acting as a rabble-rouser.

Transnational region
In 1965 Malaysian's Prime Minister Tunku Abdul Rahman urged his Singaporean colleague Lee Kuan Yew to leave the Malaysian federation. As he phrased the problem: 'there have been so many differences with the Singaporean government… It appeared that as soon one issue was solved another

---

[23] Daiva Stasiulis & Darryl Ross, 2006, 'Security, flexible sovereignty and the perils of multiple citizenship'.
[24] Ibid., p. 345.

cropped up'.[25] The basic grudge was against the political activism of Lee and his PAP (People's Action Party) in Malaysia. The fear of Singapore-backed Chinese dominance over Malaysia (Singapore, with a 90 percent Chinese population, allegedly influenced by communist China) was also an important factor. Singapore had already been labelled as Malaysian's future New York, but was now obliged to realize the vision of a world city on its own.

In view of its small size (2 million people on 630 km² in 1965) and previous commitment to a greater Malaysian federation, the sense of crisis that seized Singaporean leaders can be understood. As Ross Worthington remarks, 'The government felt surrounded by enemies and, within this context, an *ideology of survivalism* was seized upon as a mobilizing concept for the economy, society and politics'.[26] It was a program that underlined pragmatic decisions, discipline and sacrifice. Apart from a strong one-way form of authoritarian government, it also implied the instilment of values that would avert ethnic tensions. These were the 4Ms: multiracialism, multilingualism, multiculturalism and multireligiosity. Kong and Yeoh suggest that a fifth M should be added, 'meritocracy': the admission of people into the ruling elite on the basis of hard work and economic achievement.[27]

The 'ideology of survivalism' has deeply marked Singaporean politics well into the current period, and apparently irrespective of economic success. It has produced a 'hegemonic state' in which the core executive (about fifty to sixty people) makes all basic decisions and discontent is effectively neutralized by early elimination of its sources. The PAP has remained by far the dominant party, consistently winning more than 60 percent of the votes since 1965 (93 percent in 2015). A truly hegemonic system exerts its coercion always within a context of broad support; censorship is primarily self-censorship! This notion is also compatible with the new emphasis on 'Asian values', particularly the communitarian ideal of putting the nation before self. In a situation where the goal of economic survival has lost its urgency, Asian values evoke the idea that there is still a struggle for cultural survival amidst overwhelming Western influences. This should counteract the state's (essentially Western) modern project.

---

[25] The Straits Times, Augustus 9, 1965.
[26] Ross Worthington, 2003, *Governance in Singapore*, p. 39.
[27] Lily Kong and Brenda S.A. Yeoh, 2003, *The Politics of Landscapes in Singapore. Construction of 'Nation'*, p. 36.

While pursuing hegemony in the current world, one usually benefits from a technocratic context or a 'high-modernist ideology' in which goals (expansion) are uncontested and the search for means can be left to scientific researchers or business corporations. How does explicit recognition of values (identities) in the Singaporean self-definition as a multicultural society go along with a hegemonic technocratic discourse? Apart from the geopolitical message to the neighbouring countries that Singapore was not a 'hostile' Chinese enclave in a Malayan world, this definition helped the state reject special ethnic claims with the argument that they violated the principle of multiracialism and multiculturalism. In this way, the state could carve out 'a space of autonomy for itself untouched by 'racial' claims'.[28]

Despite emphasis on 'Asian' values, such as the family and the community, the Singaporean state entertains an awkward link between governance and identity. Where (Chinese) identity is obvious it must be suppressed, and where social cohesion appears it is consciously grafted into economic success rather than cultural distinction. One dimension remains paramount in Singaporean territoriality: governance. This may be seen as an expression of Asian governance, but also as a consequence of Singapore's geopolitical predicament. Although Singapore is a state lacking space, it is so unique when contrasted with other small states that it can hardly be considered a model solution for the problems of small countries. Even the Renaissance city-state of Venice had much larger territorial resources at its disposal. But Singapore's coping with the dimension of closure may be instructive because it shows the results of power concentration in the executive as well as extra-territorial actions motivated by feelings of crisis and emergency.

The first act by Deputy Defence Minister Rajaratnam in 1972 was to proclaim the vision of Singapore as a future 'global city', with the entire world, rather than just Southeast Asia, as its hinterland. When the globalizing aim reaped success, Singapore could somewhat relax its suspicious attitude toward the neighbours and more constructively cooperate in such intergovernmental frameworks as ASEAN and the Asian Regional Forum (ARF). On the local level, Singapore, Malaysia and Indonesia agreed in 1989 to facilitate transboundary movements in order to create a (SMI) Growth Triangle, which included Singapore, the Malaysian state of Johor and the Indonesian province of the Riau islands (figure 5.1). This offered a solution for Singapore's problem of accommodating further industrial expansion and finding

---

[28] Ibid., pp. 35-36.

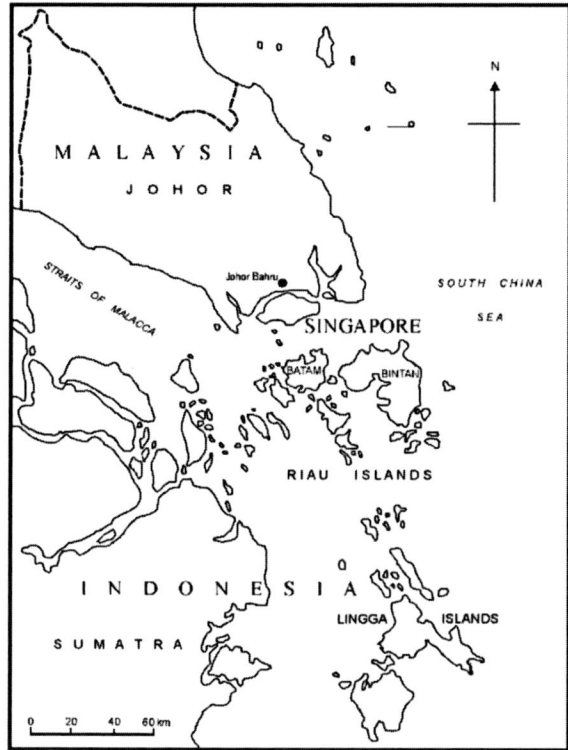

Figure 5.1. Singapore and the growth triangle area (SMI)

leisure space. The Indonesian island of Bintan was chosen for the development of an industrial estate, an extension of Singapore's manufacturing space, and an international beach resort. On the island of Batam a second industrial park arose with extensive housing accommodation. A direct ferry link between both islands and Singapore also offered the possibility of daily commuting. These developments attracted much attention among researchers interested in re-territorialization of the state and among those who proclaimed a future 'borderless world'.[29]

At face value this type of cross-boundary cooperation is similar to the creation of regional arenas in the European system of governance, but the administrative units involved in the SMI triangle are mutually quite dissimilar:

---

[29] Kenneth Ohmae, 1990, *The Borderless World. Power and Strategy in the Global Marketplace.*

a sovereign state, a federal sub-state and a province. This already suggests a power configuration that differs from the symmetric and grassroots organization that characterizes European inter-regional politics. The SMI triangle and other Southeast Asian growth 'triangles' involve 'interdependent' borderlands rather than the 'integrated' borderlands as envisaged in Europe.[30] This means that there is cooperation between Singapore, Malaysia and Indonesia without real institutional innovation like the creation of a regional transborder *authority*. As a consequence of the economic benefits it brings to Indonesia and Singapore, this (sub)regional cooperation is instrumental in lessening political tensions between the countries without causing any great loss of political sovereignty. Again this aim is quite irrelevant to cross-boundary cooperation in the European Union.

Grundy-Warr et al. conclude that the 'key to understanding developments along the Singapore-Indonesian border zone is the wish to use the locality as a resource for strengthening national sovereignty and thereby to avoid institutional innovations that weaken national controls'.[31] The tendency to implement economic development by means of self-contained projects like industrial parks or recreational resorts rather than through integrated spatial development is clear evidence of this national-level rationality. The drawback of this political configuration, however, is utter disruption of local communities, both by the rise of fences and boundaries that accompany the 'foreign' intrusions and by the presence of Indonesian migrants that are attracted to the area in the hope of using the Singapore connection for illegally entering Malaysia. Most jobs that are created in the new resorts or estates are beyond the reach of most villagers. In order to make the resorts 'world-class', employees are usually recruited from resorts in other tourist centres such as Bali.[32] As a result, extensive squatter areas have emerged that have to be cleared regularly to make space for development in one of the 'flagship' projects. In order to escape eviction, squatters have invaded the forested areas in Batam, where they caused significant environmental degradation including dangerous landslides.[33] Such developments illustrate the frontiers that global cities

---

[30] Carl Grundy-Warr, Karen Peachey and Martin Perry, 1999, 'Fragmented integration in the Singapore-Indonesian border zone: Southeast Asia's "Growth Triangle" against the global economy.
[31] Ibid., p. 321.
[32] Tim Bunnelll, Hamzah B. Muzaini and James D. Sidaway, 2006, 'Global city frontiers: Singapore's hinterland and the contested socio-political geographies of Bintan, Indonesia'.
[33] Carl Grundy-Warr et.al., 1999, p. 320.

produce, particularly when their sphere of influence is an area of 'fragmented integration' where national sovereignty is a holy principle.

The particular size and history of Singapore makes the state nervous about any infringement on the principle of sovereignty, even if it is only symbolic. This means two things: no alliance with great powers and no ethnic identification with nations abroad. In order to promote the regional balance of power, Singapore invites participation of extra-regional powers like the US, but also struggles with its interventionist practices. Singapore supported the intervention in Iraq (2003) notwithstanding their strong preference for multilateralism. Singapore welcomed the political changes in China as it offered new economic opportunities, but also had to counterpoise China's designation of Singapore as part of the Chinese diaspora with statements about Singapore's Southeast Asian identity. While Singapore's economic strength and foreign politics created a peaceful condition that was deemed utopian in 1965, this condition has been maintained by opening the country to foreign investment and influence while also preventing the adoption any official identity (for example religion) and denying admittance to any discussion or object (like the row about Rushdie's 'Satanic Verses') that may offend some international player.

National illegible regimes (E)

Shadow state

When the Soviet State actually disappeared, international practice assigned the Russian Federation as its legitimate successor, which automatically relegated the other territorial remnants to the status of secessionists. This means that important assets of the Soviet state, such as the military infrastructure and its control, passed to the Russian Federation (although troops remained on the territories of the separated states). The post-Soviet space was also characterized by the new phenomenon of diaspora. On the eve of the Union's dissolution, a quarter of all Soviet citizens lived outside their titular administrative region, particularly ethnic Russians. This transnational ethnic geography inevitably influenced foreign policies in a way that differed from traditional state-to-state politics.[34] The resulting secession may have satisfied the

---

[34] Charles King and Neil J. Melvin, 1999/2000, 'Diaspora politics: ethnic linkages, foreign policy and security in Eurasia'.

longing for independence, but it did not reinstate the prototype of the sovereign state. In such cases, the new state will either remain uneasily tied to the legitimate successor state or alternatively be cut off from former human and material resources. Secession meant freedom, but it also facilitated cross-border strategies of organized crime. Boundaries hid illicit activities from view and the perpetrators could benefit from Russian as lingua franca to get easy access to the 'near abroad'.

While this situation applied in varying degrees to all the former Soviet republics, Soviet nationalities politics seems to have prepared those very republics for an independent existence, at least in terms of a government of their own. Yet, this did not guarantee a smooth transition to well-integrated states. All ex-Soviet republics had to cope with varying degrees of internal disorder, and in some cases the new government was a thinly veiled autocratic regime parading as a democracy (a so-called 'virtual democracy'). Andrew Wilson has pointed to the skills of post-Soviet political leaders to fake democracy, either by purposively staging extreme political forces like the nationalist Zhirinovsky in Russia (and presenting themselves as the 'lesser evil') or by creating new parties in a divide-and-rule strategy (or if that does not work simply by straightforward meddling with electoral results).[35]

At the collapse of the Union, the idea of regulation by means of law was far from everyone's mind. In the Russian federation, this meant that safeguarding individual freedom took precedence whatever its social costs. This attitude was reinforced by the interpretation of the transition from communism to capitalism as an imperative to privatize everything that was historically always claimed by the state. One of the nasty results was the 'capture' of the state for personal gains, corruption or the misappropriation of government funds. Two stories that caused a great deal of controversy were the Magnitsky affair in 2009 and the allegations about links between Russian government and the underworld in the Wikileaks diplomatic cables.

> **Box 5.3. Magnitsky case**
>
> In 2007, officers from the Moscow Tax Crimes Department of the Interior Ministry raided the offices of the Hermitage Capital hedge fund and its associated law firm Firestone Duncan. As William Browder, founder and CEO of Hermitage Capital, later testified: 'They were particularly intent on getting hold of the statutory documents of our investment hold-

---

[35] Andrew Wilson, 2005, *Virtual Politics: Faking Democracy in the Post-Soviet World*.

> ing companies – the seals, charters, articles of association of our investment holding companies. They seized all of those documents even though they had nothing to do with the pretext of their search'.[36] A few months later, the firm was notified by the court of St. Petersburg about judgments that had been issued against these holding companies running into a couple hundred million dollars. Lawyer Sergei Magnitsky, partner of the law firm Firestone Duncan was asked to investigate the background of these claims.
>
> Magnitsky discovered that the said holding companies had been re-registered into the name of another person (a convicted murderer) and that the new 'owners' had applied for a refund of the 230 million dollar tax that had been conscientiously paid in 2006. This sum had subsequently been collected by the fraudulent group, which included police officers. Magnitsky lodged a complaint against those concerned (June 2008), but the police officers named in the criminal complaints now opened a new criminal case against the lawyers working for Hermitage Capital. Magnitsky, convinced that he was in the right, did not want to flee from the country (as he was advised to do) and had to face pre-trial detention. In prison his health deteriorated while medical treatment was withheld. On November 16, 2009 he died. As Browder notes, 'He entered prison a healthy 36-year-old man and eleven and a half months later he was dead'.

Indignation in the US Senate about the case led to the Magnitsky Act of 12-12-2012, which denies entrance into to the US for anyone judged responsible for Magnitsky's death. In retaliation, the Russian government put a moratorium on Americans adopting Russian children and issued a list of Americans that were forbidden from entering the Russian Federation.

In congruence with and following the Magnitsky case, governments and law enforcers elsewhere in the world do not dismiss reports about links between criminal groups and the Russian government as isolated incidents. At least that much is suggested by the Wikileaks publication of US embassy cables. One of these telegrams, sent in February 2010, contains information obtained from Spain's national court prosecutor, José Grinda Gonzalez, who was engaged in rounding up a Russian criminal gang in Spain. The prosecutor suggested that 'the Russian mafia… exercises "tremendous control" over certain sectors of the economy, such as aluminium'. He also said 'to believe that whereas terrorists aim to substitute the essence of the state itself, OC [orga-

---

[36] William Browder, 2009, *Hermitage Capital, the Russian State and the Case of Sergei Magnitsky*.

nized crime] seeks to be a complement to state structures'. He summarized his views by asserting that the GOR's [Russian government] strategy is 'to use OC groups to do whatever the GOR cannot acceptably do as government.' Political scientists have also brought to light the strongly 'informal' nature of activities performed by officials in the post-Soviet states. One of them, Johan Engvall, calls the post-Soviet state an 'investment market'.[37] A core feature of this system is the practice of paying out for a job in the police, tax-administration or other governmental service, but also for entering the parliament or a course in higher education. The custom does not simply boil down to the worldwide practice of paying bribes, but is predicated on getting sustained access to revenues entirely apart from the salary that is attached to the job. For example, to secure a place as governor or on the Federation Council of the Russian Federation, one had to pay between 5 and 7 million dollars. In Kyrgyzstan, members paid 300 thousand dollars to enter the parliament.[38] Little imagination is needed to conclude that recapturing one's investment and gaining a sustained profit from the job involves practices that cannot be tolerated in a state upheld by the rule of law. Although one may still consider such conditions as growing pains of a state recently liberated from totalitarianism, problems that will be finally subdued under 'a firm leadership like Putin's', an analysis of financial and governance indicators suggests a continuing gap between many of the post-Soviet states and the major developed countries of the West.

A report by the Global Financial Integrity group estimates the illicit financial flows (resulting from deliberate misinvoicing) in and out of Russia between 1994 and 2011 as about 30% of the total recorded in- and outflows.[39] Such illicit flows are strongly indicative of the size of the underground economy, which between 1991 and 2015 is estimated on average 38% of GDP. The last ten years of the period show a somewhat lower average of 33%. A comparison of the size of the underground economy in the G7 countries[40] and Russia reveals the wide gap even with Italy, the notorious heartland of the Mafia (Table 5.1).

---

[37] Johan Engvall, 2015, 'The state as investment market: a framework for interpreting the Post-Soviet state in Eurasia'.
[38] Ibid., p. 31.
[39] Dev Kar and Sarah Freitas, 2013, *Russia: illicit financial flows and the role of the underground economy*.
[40] Canada, France, Germany, Italy, Japan, United Kingdom, United States.

Table 5.1. Size of the Shadow Economy in G7, Georgia, Ukraine, Italy, Russian Federation, China, and the world (158 countries): 2015 and average 1991-2015 (% of GDP). *Source:* Leandro Medina and Friedrich Schneider 2018

	2015	Average 1991-2015
G7	10.8	13.5
Georgia	53.1	64.6
Ukraine	42.9	44.8
Italy	23.0	25.0
Russian Fed.	33.7	38.4
China	12.1	14.7
158 countries	27.8	31.8

While the size of the Shadow Economy can be considered an indicator of the strength or the weakness of the state, there are other indicators that more directly address differences in quality of governance. The World Bank collects data about specific dimensions of governance, such as voice and accountability, political stability, government effectiveness, regulatory quality, rule of law and control of corruption. Such indicators are based on perceptions which although 'subjective' have the advantage of measuring the reality "on the ground" rather than *de jure*. Differences between occupational perceptions are overcome by tapping the perceptions of many different groups (occupations).[41] All these indicators show a sustained gap between the post-Soviet countries and the G7 countries in the period 1996-2016 (figures 5.2 – 5.4). China matches the post-Soviet pattern rather than that of the G7 countries. On one indicator, 'voice and accountability'[42], the gap between post-Soviet states and the G7 has become even wider (figure 5.4). Only one country, Georgia, shows signs of escaping this fate after the power shift to president Saakashvili (2004), a change that echoes its foreign policy orientation towards the West.

---

[41] Other types of systematic bias, for example influence of the level of economic development on judgments about governance, have been tested and found insignificant. Daniel Kaufmann, Aart Kraay and Massimo Mastruzzi, 2010, *The Worldwide Governance Indicators. Methodological and Analytical Issues.*

[42] 'Voice and accountability' involves (perceptions) of the extent to which a country's citizens are able to participate in selecting their government, as well as freedom of expression, freedom of association, and a free media.

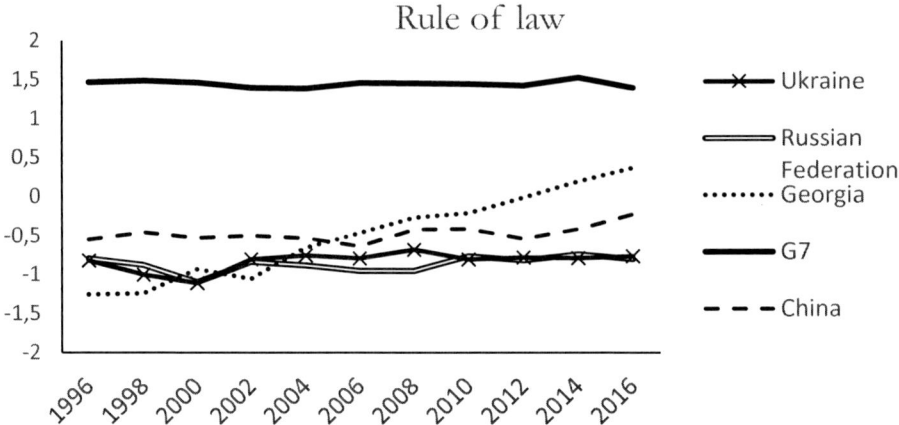

Figure 5.2. Quality of governance: Rule of law 1996-2016. Source data: World Bank (data.worldbank.org), Worldwide Governance Indicators.

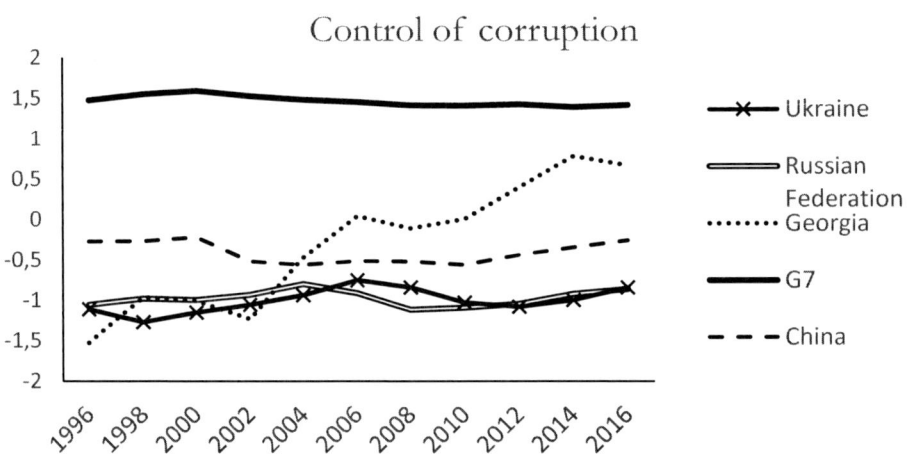

Figure 5.3. Quality of governance: Control of Corruption 1996-2016. Source data: World Bank (data.worldbank.org), Worldwide Governance Indicators.

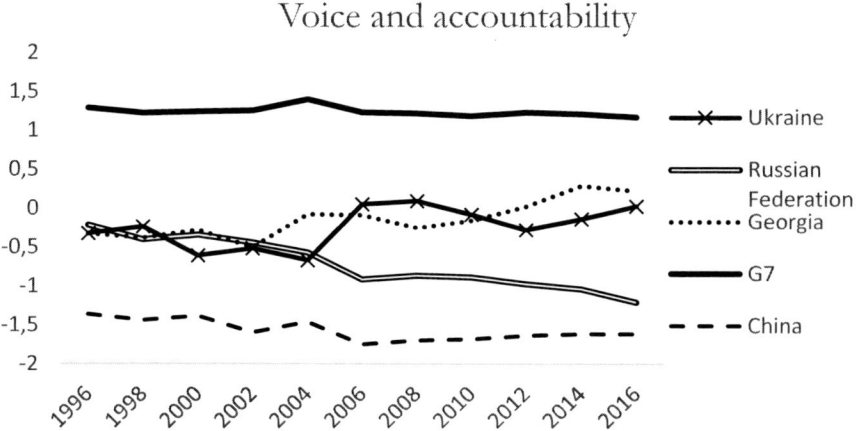

Figure 5.4. Quality of governance: Voice and accountability 1996-2016. Source data: World Bank (data.worldbank.org), Worldwide Governance Indicators.

These indicators do not entirely mirror the differences in the size of the shadow economy, which is also a function of the labour market and the presence of large or stable enterprises. The slight worldwide shrinkage of the shadow economy during the last two decades also confirms its inverse relationship with economic development (see also China's score). The tenacity of the 'quality of governance' gap in the graphs from figures 5.2 – 5.4, however, suggests an enduring state strength that differs from the liberal system of governance. It rests on a strong executive who spends extensively on military and security resources while simultaneously being sustained by a shady system of influence and benefits (often called 'oligarchic'). At the same time, the state directs public opinion with strong ideas about good and evil both in terms of social values and differences between countries or continents. It remains to be seen whether the G7 states will remain immune to this type of illiberal governance, or if they will succumb (a downfall that has been already speculated as a possibility ('friendly fascism') for the United States by a visionary author in 1980).[43]

---

[43] Bertram M. Gross, 1980, *Friendly Fascism: The New Face of Power in America*.

Private illegible networks (G,H)

National networks: Sub-Saharan Africa

If preserving state sovereignty is the guiding principle for the international community while addressing a country's domestic affairs, Africa is a nightmare for any UN official. The dominance of informal networks and non-state agencies that perform quasi-government functions reminds us of the fact that African states have never actually satisfied the Westphalian (or Weberian) ideal. Two reasons for this are: 1. The continuance of tribal and clientelistic traditions, and 2. Extreme poverty that prevents the accumulation of substantial government revenues. While poverty on the one hand produces insensitivity to administrative regulation, the government on the other hand tries to secure external sources of income (donors) or a share in the shadow economy (40% for sub-Saharan Africa between 1996 and 2015). This means that there is often a lack of regulatory authority; sources of power remain obscure and are often personal. Building a network of clientele is more important than territorial governance, which means that areas of influence can easily extend across state boundaries. An example is the control exerted by Charles Taylor's rebel militia, the National Patriotic Liberation Front, which ruled a large part of Liberia and parts of the neighbouring states between 1990 and 1996 ('Taylorland').

Governance in the African state is the product of many intersecting forces: international non-governmental organizations (INGOs), multinational corporations (MNCs), local non-state powers (warlords or community leaders) and state officials that actually rely on informal networks and shadow economies. Many (international) actors, compelled by necessity, provide 'governance' independently from the state in which they operate. An example is the oil industry that, in response to local protests, has been engaged in community development projects like housing and environmental care (for example, Shell in Nigeria). The hiring of private security by MNCs is a well-known practice in Africa that often corresponds with government responsibility in other parts of the world. Additionally, African rulers have been deeply involved in economic networks for personal economic gain, including accepting bribes from multinational firms (shadow state).

While transnational actors are providing quasi-forms of governance, they may also introduce divisiveness and feed the shadow economy. Specific economic activities that can be easily individualized, like gem mining, are notorious for their capacity to undermine the control of agencies promoting public interest (local, state or INGO). The value of gems entirely depends on the global market and on global networks of gem dealers, but state officials may also constitute an essential part of the network. Rosaleen Duffy aptly analysed the 'shadow state' involved in illicit sapphire mining in Madagascar:

> Economically impoverished Malagasy people travel to the gem areas to seek employment; they either work alone or are organized through clandestine networks headed by individuals in the Malagasy elite. The diggers then sell their unpolished stones to gem dealers… Finally the gem buyers traffic the stones out of Madagascar through airports or by sea, with assistance of key individuals within relevant Customs departments, government agencies and local business… Despite …attempts at governance…the gem sector has still largely remained beyond control. In order to understand this it is important to examine the logic of gem rushes and see how the sector developed into its current form [44]

The 'logic of gem rushes' pivots on the prospect of fortunes made in a single day and the haphazard method of mining that results from this outlook. For one thing, diggers' actions are often ill directed due to their being based on rumours rather than on reliable information. Attempts to protect the environment by delineating no-go areas made things even worse in Madagascar, because these very actions were interpreted as a sign that the protected area contains the most precious and heavy stones. This is the paranoid logic inherent to systems in which promotion of the public interest has become a curiosity.

Notwithstanding the 'Hobbesian' paradigm of self-interest that seems to be characteristic to the shadow state, the exchange of valuable goods, services and money is in itself a token of order. Although asking payments for privileged *access* to the market or to services is a gross denial of the ideal of the regulatory state, it evokes a similar sphere of predictability. Payments are at least an understandable principle that offers a personal escape that a weak state or violent regime itself does not provide.[45] Even further, the system in Chad and northern Cameroon may even yield some redistribution, like the financing of mosques and Muslim schools by wealthy individuals.[46] The networks involved in this process of accumulation do not necessarily coincide

---

[44] Rosaleen Duffy, 2005, 'Global environmental governance and the challenge of shadow states: the impact of illicit sapphire mining in Madagascar'.
[45] Janet Roitman, 2001, 'New sovereigns? Regulatory authority in the Chad basin'.
[46] Ibid., p. 250.

with state territory – they even compete with the state's appropriation of wealth – but the established state may nevertheless benefit from it because these networks supplement the starvation wages of civil servants and pump hard currency into the economy. To put it briefly: new systems of accumulation are essential mediations between the state and the world economy. Sovereignty as a principle is not under attack; what changes is the geography of power and the regulatory means of the state.

While the 'logistical' benefits the state inadvertently reaps from an informal economy are substantial, international legal status simultaneously remains an important asset for non-state actors. One of the reasons for this is that governments may sue multinational firms in European or US courts for doing business with insurgents.[47] Even insurgents like Zaire's Laurent Kabila, says William Reno, 'hastened to Kinshasa to assume the mantle of international recognition, rather than exploiting his standing in a trading area that had its own currency and far more cultural ties to eastern neighbours than to the distant capital'.[48] This, however, does not mean that the quest for sovereignty among such local strongmen also includes the ambition to provide public goods.

Because rulers' legitimacy in weak states is always under attack, they need all resources for the 'urgent task of managing potential rivals'[49]. Spending money on public goods like health care or education appears as a waste of money in an emergency state. It can be extremely difficult to escape the lure of using the treasury for power strategies, as was shown in Kenya in 2002. The new government came into power under the general expectation that they would end the previous government's tradition of corruption, which had shied away from most international donors. The new regime even established institutions to inspire confidence, like the Kenya Anti-Corruption Committee, and instated a 'permanent secretary for ethics and governance', John Githongo. He was not granted a quiet job. Reports soon reached anti-corruption officials that a non-existent company, Anglo Leasing, had been awarded a very lucrative contract (55 million USD) with the purpose of building a forensics lab for the Central Intelligence Department. It never was built. A list surfaced of other companies and representatives without addresses that

---

[47] William Reno, 2001, 'How sovereignty matters: international markets and the political economy of local politics in weak states'.
[48] Ibid., p. 202.
[49] William Reno, 2005, 'The politics of violent opposition in collapsing states', p. 133.

had been awarded contracts (in total 600 million USD, 19 percent of the state budget) for activities mostly in the sphere of security, policing, satellite services, etc.[50] Asked for an explanation about the invisible results of these firms, the 'Office of the President said that security suppliers operate mysteriously behind the shadows'.[51] But it soon became clear that senior government members were involved in these operations, mainly, as they declared to Githongo, as resource mobilization for political parties in power (the National Rainbow Coalition). Once he penetrated to the core of the affair, Githongo was implored to keep quiet in the interest of national stability. After receiving death threats, he decided to take refuge in England where he made public his report to Kenyan President Kibaki. Notwithstanding the shocking image of corruption Kenya now had in the highest circles, this story has a silver lining. The less eye-catching and untold details are of the people who exposed the corruption: those specifically appointed to stop corruption, local newspapers writing about the issue and Kenyan members of parliament who became alarmed and did not stay silent. International money often brings more opportunities for (illegal) personal enrichment, but it comes with strings attached: international norms of 'good governance'.

In Kenya, central state representatives dominate the scene of corruption. Contrastingly, in collapsing states former elites may lose official power, but they often continue using their networks to lay claim on the shadow economy. The collapsing state may show two types of disintegrating forces: former elites that indulge in predatory behaviour and formerly marginalized elites that pursue a new independent power base (territory). Reno suggests that armed groups formed around local elites formerly linked to the (ousted) central authorities are more violent and predatory than armed groups who form around marginalized elites.[52] In the latter situation, new leaders garner support from a particular cultural group (e.g. Somaliland, the Yoruba in Nigeria) by articulating a political program and the supplying a small selection of public goods, such as security and redistribution of resources. When externally supported local politicians succeed in controlling the bulk of material resources, looting behaviour by the disadvantaged is often the unfortunate result. In order to suppress such rebellious behaviour, local elites bring in

---

[50] John Githongo, Report on the findings of graft in the Government of Kenya, to H.E. President Mwai Kibaki (summary document), 22 November 2005.
[51] Ibid., p. 15.
[52] William Reno, 2005, pp. 128-131.

vigilante groups that employ cruel methods of punishment. The drama of state collapse arises when the remnants of an integral territorial government (who still attach some value to the state) become predatory and act to further disintegrate society by introducing violence. Consequently, leaders that defend the provision of collective services are only able to do this by calling for the dissolution of the state.

While discussing shadow economies and corruption, one always bumps into examples from Africa. But the loosening of state control and the erosion of public values has also become more prominent in the US and Europe. There are undoubtedly more institutional public and private watchdogs in the West, but the freedom of parties or presidents to grant important contracts to firms that support the ruling elite (George W. Bush administration and Iraq reconstruction contracts) or to dominate the media (Berlusconi in Italy, Trump in the US) is worrying. Granting more power to the executive is even accepted as a means to get rid of the paralysing game of particularistic interests. While these phenomena are not a sign of state fragmentation, they may undermine a state's legitimacy in the long run. Fragmentation of control in the West is more often the result of groups in civil society possessing a knowledge advantage than the predatory behaviour of political elites. The strength of the state in Late Modernity no longer depends on mobilization of an ethnic nation, but rather on civil networks and social media. Such networks, however, come with unique limits to their control as well.

Transnational network: Islamic terrorist networks
The terrorist networks we have come to know since the turn of the century (Al Qaeda, Taliban, Al Shabab, Boko Haram) usually have a (secret) territorial base, but there is only one that claimed to possess a state ('caliphate'): the Islamic State. It owed its existence to the collapse of states in the Middle East, but its ambition was much wider than accepting the provision of a safe haven. The territorial base was simply to be a seed for a worldwide revolution to make Islam the universal religion. The aim is not only to expand the territory of the Islamic State over the neighbouring countries of the Levant (*Sham*), but also to arouse conflict between Muslims and non-Muslims in the rest of the world (particularly in Europe with its substantial Muslim population). Spreading discord and fear is its central strategy and the possession of a territorial base constitutes an important strategic asset in this pursuit. Territory made IS's picture of the future more realistic and exciting for potential fol-

lowers (jihadists) in the West and it provided a base to produce propaganda and material for terrorist attacks. Of course, this also meant that they became an easier target for military attacks, but the presence of a civilian population (de facto hostages) promised some protection.

The most intriguing phenomenon to Western observers is the successful recruitment of young people, who until recently had led a peaceful existence with their families in Europe and never showed any radical tendencies. In the words of a jihadist, 'I experienced a void in my life, a lack of meaning'. A sociological analysis would conclude that young people in certain social and urban environments receive too few challenges and possibilities to develop their skills and lack the opportunity to make a meaningful contribution to society. A frequently encountered scenario is that a youth commits a petty crime, ends up in jail, and gets radicalised by inmates preaching a religious message of redemption. Once on the path towards a new spiritual fulfilment, it is easy to encounter omens in social media and websites that confirm one's direction and show the way to the IS utopia.[53]

IS exploits the perverse side of the young, mobile and urban professionals of late modern society. The Western tendency to externalize blame for this phenomenon, casting it as a product of a collapsed society or of poverty, ignores how much the movement owes to the West. Indeed, the mental products of globalization, such as violent computer games, cosmic wars and heroism, arguably laid the groundwork for radicalization in many young minds. IS is far from a product grown in areas untouched by globalization, it is globalization itself.[54]

Conclusion: transcending the Westphalian state?

Despite lamentations about the center that 'cannot hold', the baseline of this chapter is that many actors tend to cling to the territorial state and to the principle of sovereignty even under disorderly circumstances. This is not to deny that institutional and material changes erode the hierarchical and closed territorial ideal of the Westphalian state. Since the last decades of the 20th century, state governments or 'executives' increasingly rely on internal actors – cities, regions or private organizations – for (foreign) policy making (glo-

---

[53] Gilles Kepel and Antoine Jardin, 2015, *Terreur dans l'Hexagone. Genèse du Djihad Français*.
[54] Ahmed Al-Rawi, 2016, 'Video games, terrorism, and ISIS's Jihad 3.0'.

calization). On the other hand, they concurrently interfere more and more with security problems, economic projects or directly with the citizens in other countries (transnational regimes). Even more confusing is the informality that has penetrated political domains, to the extent that in certain cases the terms shadow state and shadow economy apply. One may conclude that the traditional territorial shell with its system of public control has become rather hollow, but there are still gradations on the scale, with the global North on one end of the range and the global South (Africa in particular) on the other. Even intergovernmental security arrangements in the North have implied that governments do not merely exercise influence *on* other states but also *in* other states.

Many examples of non-legible or informal practice discussed in the previous pages reveal a passionate attachment to territorial independence and international recognition. But at the same time there is a tendency to let boundaries grow more and more porous. This resembles a return to the ancient imperial territorial order where borders could only be defended by frequent raids in barbarian lands or by creating vassal or client states as buffer zones. The comparison cuts some ice when we consider the US security policy against terrorism or the Russian attempts to return regions with a (pro)-Russian majority in former Soviet Republics (Ukraine, Georgia) to the core. In other cases, the practice results from the mere need for economic survival in a globalized world (Singapore), or from the opportunity to benefit from high profit yielding products like gems or precious metals in a country that creates few products with much added value (as is the case in numerous African countries). The imperial model of territoriality assigns an *active* role to closure and governance, which is also the case for state executives in the globalized territorial order. But what about the *passive* quality of *identity* associated with the imperial order?

Our time feels overloaded with identity politics: some groups demand recognition of their sexual identity, others claim the religious identity of a state and others again want to worship national tradition. Much of this is either non-territorial or a *reaction* to globalization. There is no sustainable active principle of building territorial identities in the late-modern territorial order (table 5.2) like the one that prevailed in the high-modern period. On the contrary, manifestations of ethnic identity (even of majorities like the Chinese in Singapore) are sometimes suppressed as being counterproductive to the global strategies of a state. What we experience, however, is a threat from the

Chapter 5. CAN THE CENTRE HOLD?

foreign 'other' as someone who either denies our values or wants to exploit us by entering our territory. We feel suddenly exposed now that the passive assumption of *closure* (passed down from the previous era) does not seem to work anymore and needs reactivation, if necessary in the shape of a border wall that is more symbolic than real.[55] This is one side of the territorial shock of our time, in which the shift to passive identity constitutes the other side. Where identity in the past was - at least partly – supplied by the nation-state, now people must choose from a conglomerate of border crossing movements and ideals. This has immediate repercussions on the public trust in democracy. The feeling of not being represented may end in the espousal of populist movements (United States, India, Eastern members of the EU) or an autocratic regime if it can supply material success (China, Singapore).

There is reason to expect a period of geopolitical disorder to emanate from the political reactions to current territorial shock. States may leave supranational organizations like the EU or cancel international agreements on trade or peace (US-Iran). Since populist politics is mostly rhetoric, results will be disappointing or worse: it may even mean war when the symbolic creation of an external enemy becomes reality. This geopolitical stage does not change the territoriality of the current era in which the intensification of transborder communication cannot be reversed (just as the free organization of people around common goals cannot be endlessly repressed in China (page 140). As the history of the Chinese Wall (Chapter 2) shows, walls cannot be made impermeable and they may be a signal of a deep flaw in society rather than a restoration of state power. It may take some time for voters to become aware of this reality, and meanwhile some dramatic events will take place: perhaps the collapse of great powers (contemporary 'empires') and the worsening of climate conditions. Our territorial shock cannot be quelled by resurrecting the territorial order of a past era. What we need is a moral principle that helps us face new conditions with confidence.

One mode of accommodation is glocalization, an intense involvement with a small world, small enough to inspire confidence but at the same time reflect the world in its visitors, exhibitions and enterprises. It should be geared toward participation and actively meeting or collaborating with foreigners. Another suggestion is a world of communities based on faith. One

---

[55] Reece Jones, 2012, *Border Walls: Security and the War on Terror in the United States, India and Israel.*

author described this new vision that deviates so essentially from the mainstream conception of international relations: 'a "corporate association" of peoples and nations that are bound together by the flow of ideas and practices embodied in religions, customs and traditions rather than abstract rights and commercial contracts'.[56] While the latter portion of this vision dangerously invokes a world marked by clashes of civilizations, as a whole it at least recognizes the integrating force of values. But this value and moral imperative should be conquering the common threat of global environmental collapse rather than religion or custom. None of this advocates the abolishing of state boundaries, which remain practical in allocating resources and maintaining infrastructural accomplishments.

Table 5.2 Late-modern territoriality

Dimension	Active/passive	(Un)boundedness
Closure Structure of authority. Who are (un)affected by dominant authority?	Multi-level coordination (Active)	Glocalization Porous boundaries
Governance How is a territory / group engaged in turning it into a resource?	Institutional proliferation (Active)	Controlling resources (corruption etc.) Global finance and trade
Identity How do people attribute *meaning* to power in (territorial) assemblages?	Corporate associations of people and nations (Passive)	Religion Civilization

While one considers the appearance of walls and neo-nationalism on the one hand and the many identity movements springing up as moral adaptations to a new territoriality on the other, the actual state histories, the reality of the situation, recedes from view. To arrive at a sharper picture, we should not merely list new movements and powers that populate the scene, but search for cases of a *hard reset*. The term is derived from the world of information technology. All intelligent hardware of our time occasionally suffers from getting stuck, a condition that can be overcome by a hard reset. This eliminates a running program or erases part of the memory (to the sorrow of the user) but opens the system for new information and codes. The territorial parallel of the hard reset is the dying state.

---

[56] Adrian Pabst, 2012, 'The secularism of post-secularity: religion, realism and the revival of grand theory in IR', p. 1017.

Chapter 6
DYING STATES:
PRELUDE TO RE-TERRITORIALIZATION?

> It is precisely in the wake of great shocks, in the seclusion of closed perspectives, that thought, suddenly thrown upon its own resources, regains great hope and designs new projects. Mankind is engaged in a battle, in a massive action. Let us take advantage of the reprieve that has been given to us…
>
> **Joseph Rovan,** French historian writing in occupied France, 1941 [1]

> (Deterritorialization) [D] is the movement by which 'one' leaves the territory. It is the operation of the line of flight. There are very different cases. D may be overlaid by a compensatory reterritorialization obstructing the line of flight… Another case is…when [D] prevails over the reterritorializations, which play only a secondary role, but nevertheless remains *relative* because the line of flight it draws is segmented, …sinks into black holes, or even ends up in a generalised black hole (catastrophe).
>
> **Gilles Deleuze & Felix Guattari,** *Mille Plateaux,* 1988 [2]

The line of flight and the music of our time

Poland in the 18th century was a country with vast resources in terms of territory and population. It could even boast of such 'modern' institutions as a parliament, the *sejm*. Yet, the capability to turn these resources into an effective state was practically nihil. Individual veto rights paralysed decision making in the parliament, while its members, the *szlachta* (nobility), formed rival governments, if necessary supported by militias. Polish state revenue was only one seventy-fifths of France, and the entire state budget was less than the English proceeds from postage stamps.[3] The neighbouring powers Russia, Austria and Prussia could easily take possession of this state with the pretext that it had become a risk to the security of Europe.[4] A sequence of aggressive acts that became known as the Polish partitions finally resulted in the death

---

[1] *L'Europe n'est pas seule*, 1941, quoted in Bernard Bruneteau, 2003, *'L'Europe Nouvelle' de Hitler. Une Illusion des Intellectuels de la France de Vichy*, p. 198. Author's (GD) translation.
[2] Text taken from the English translation (Brian Massumi): Gilles Deleuze & Felix Guattari, 1992, A *Thousand Plateaus. Capitalism and Schizophrenia,* p. 508.
[3] Adam Zamoyski, 1992/1997, *The Last King of Poland,* p.5.
[4] Ibid., p. 204.

of the Polish state in 1795. This did not mean however that the state, or rather the nation, was dead in the mind of its former citizens (which of course helped in the later resurrection of the state in 1919), but in international relations Poland had become a void for more than one-hundred-twenty years.

There is a striking difference in the death or collapse of states in our own epoch, where states may languish to the same degree as Poland in the 18th century or even worse and yet remain on the map. Nowadays, there is a strong international norm condemning conquest and resisting the aggressive dissolution of states. It cannot guarantee that violent occupations and secessions will fail to occur, but it acts as a deterrent against great power strategies. This produces a notable side effect: collapsed states nominally continue their existence on the world map and even show up with 'representatives' at international meetings. In reality, such states are 'black holes' in which no effective government can peacefully rule, but where external powers have free play. They suffer from the complete de-territorialization of those living on their territory, which means that the majority of the inhabitants cannot bring the state (or territory) to bear on their lives or ideals in a significant way. The ultimate manifestation of this de-territorialization is literal removal from the territory by a person's flight or migration. In the more philosophical use of the term, *line of flight* refers to any strategy to ignore or sabotage the mechanisms that pertain to a 'territory', for example by -advocating religious imperatives. Although we could ignore 'black holes' or voids in the international system in the past, the collapsing regimes in Iraq and Syria of the twenty-first century have caused shock waves all over the world, including in the self-satisfied West. Both the massive influx of asylum seekers from Syria and the violent acts of the Islamic State in Europe and America turned de-territorialization into a dramatic public spectacle. Globalization has acquired a new fearsome face.

Such developments have made thinking about re-territorialization not only a philosophical but also a practical assignment. As we have seen in the earlier chapters on epochal geopolitical earthquakes, the transformation to a new type of political map involves re-territorialization: the adoption of a new system of governance, closure and identity, which in turn demands a new mind-set in a religious, or rather spiritual, sense. Forced re-territorialization after conquest may be an uphill battle for new (foreign) rulers; it is not surprising that more than half of state deaths between 1814-2000 were impermanent : the victim was resurrected after some time as historical data collected by Tanisha Fazal have shown.[5] Resurrection after death also offers the unique opportunity to radically adapt to a new geopolitical or territorial order. States that perpetually continue their course, on the other hand, often do not adapt specifically because they do not have to cope with economic and political insecurity. The more concrete, although somewhat counterintuitive, question is: are resurrected states better equipped to take advantage of a system in which private actors (including terrorists) are sometimes stronger than states and in which territorial sovereignty is 'holey' (i.e. with *holes*) rather than holy? There is no database to underpin a sound answer to this question, but we can at least visit some cases of state death and their aftermath in recent history.

Three events inevitably promote state death: conquest, the (more or less peaceful) dissolution of a state and the collapse of effective government (often accompanied by civil war). We are going to look at one case of each: conquest (France in World War II), dissolution of an empire (Soviet Union 1989) and collapse (Somalia since 1991). The first two events have many predecessors in the history of states, but the victim state's continuation and re-territorializations may follow a road that differs from earlier territorial epochs. One outcome of the massive state-dying in Europe during the Second World War was the rise of the European Community and later the European Union. Although this result materialized years after the liberation of these countries, it was nevertheless a true break with the state ideal of High Modernity. The demise of the Soviet Union after 1989 did not only resurrect some previous states and create new ones, but also perpetuated the existence of its successor state, the Russian Federation, with the adoption of a new identity (forgoing international communist solidarity) and the creation of dependent semi-states around it. In this sense 'born again' has similarity with the (Christian) religious use of the term, which implies a rejection of one's previous identity.

The philosophers Deleuze and Guattari have capitalized on the vocabulary of territory (and other spatial concepts like rhizome or holey space) to make clear how human routines or systems of action and meaning evolve and break away ('line of flight') to find new ground. They did not in particular address political territoriality as dealt with in this book, but the ease of employability

---

[5] Tanisha Fazal, 2007, *State Death. The Politics and Geography of Conquest, Occupation and Annexation*.

of the term supports the notion that spatial events are an essential model for describing social processes, rather than the reverse. De-territorialization as 'leaving the territory' (emigration) obviously means leaving behind a set of laws that regulated one's behaviour. One may reverse the statement and omit the idea of movement in geographic space. The result is a more elementary definition of de-territorialization: renouncing laws. One may also state that leaving the territory, either literally or figuratively, describes the experience of boundary change (either on the map or in implementation). This is the meaning of de- and re-territorialization in this book.

'Line of flight' denotes the way de-territorialization proceeds. But as the quote at the head of this chapter suggests, it may also contain the seeds of **re**-territorialization. The 'line of flight' is not the premeditated strategy of a revolutionary vanguard, but rather resembles a line of fracture in a solid substance on which a certain pressure is exerted. Fault lines are obscured ahead of a crisis. The cause of the breakup of the Soviet Union is often diagnosed as nationalism, but the fractures did not follow the old ethnographic map. From its start as a communist state, the national dynamics were stimulated and manipulated by the Soviet regime. They unwittingly laid the foundations for a breakup. Moreover, an ethnic or national heritage is not something that automatically emerges in times of crisis. One needs a special resonance between current fashions and traditional roots to expect a nationalist revival. One such course of events in the Baltic republics was labelled the 'singing revolution'.

On May 14, 1988, the Tartu Pop-Music Festival staged a performance of five songs by the Estonian composer Alo Mattiisen. The event created a stir that went into history as 'the singing revolution'.[6] By repeating a brief melodic phrase from an old folksong wrapped up in the idiom of new rock music, the composer induced a shock among the audience that amounted to no less than the experience of liberation. National song and dance festivals had been a tradition in the Baltic States from the 19th century onward and continued under the Soviet regime as the only legal opportunity for large numbers of people to gather together. These manifestations meshed with the Soviet nationalities policy of supporting national cultures, although that policy certainly did not permit the singing of national anthems. However, the mobilizing effect of music was so gentle that it was difficult for the com-

---

[6] Henri Vogt, 2005, *Between Utopia and Disillusionment: A Narrative of the Political Transformation of Eastern Europe.*

munist authorities to take action, particularly since the regime was experimenting with new forms of tolerance and openness.

The feelings that were aroused did not immediately put territorial secession on the agenda. They were rather couched as a complaint about limited democratic rights and 'green' awareness of damage to the environment. In 1989 the commemoration of the fiftieth anniversary of the Molotov-Ribbentrop pact intensified the sentiment that a historic injustice had been committed against the Baltic people. This inevitably introduced the issue of independence, which had a contagious effect on other Soviet republics. This sequence of events, which ultimately led to the collapse of a state, may have remained ineffective under other circumstances. The difficulty of forecasting the future in such situations is exemplified by the fact that even in 1989 new publications in the West predicted a stable Soviet Union for decades to come. The collapse of states and regimes shares a vital feature with other complex systems: a 'tipping point'. The system can absorb significant damage before it suddenly shifts into a qualitatively different state that cannot easily be reversed.

Briefly, to know what unleashes a process of (de)territorialization, we should know both the past and the 'music' of our time. The interesting thing is that however local old injuries are, the music is unmistakably global. The lines of flight in this era, even those of the religious fundamentalists, do not necessarily point to withdrawal from the world. As the Islamic State has shown, the target is the global system and the means are territorial bases.

A temporary abode: Vichy France and its aftermath

When German armies crossed the border into France on May 10, 1940, neither the German military planning staff nor Hitler had a clear image of what should become of France in the future. The main war motive was to humiliate France and to eliminate its power once and for all. Whether this would involve the annexation of France, in whole or part, or the creation of a rump state subservient to the *Reich* was undecided. The immediate goal was clear: France should be incorporated into Germany's war economy, and France's whole military capacity, particularly as embodied by the fleet and colonial territories, should be used to shore up the attack on Britain. When Paris fell on June 14th, the defeatist leadership of the French army and state agreed to an armistice, which resulted in subdividing France into several territories with different administrative meanings (figure 6.1). For one area, Alsace and Lorraine (5), the future was without doubt. It would be annexed by Germany and would not even receive a military administration like the other occupied territories. Then there was a part of France that was not occupied, Vichy France (6). Its boundary ran roughly in an east-west direction from Geneva to Tours at the Loire and from there southward to a point in the Pyrenees about 50 kilometres inland from the Atlantic coast. It covered the entire southern half of France (save the part East of the Rhone river), 40 percent of the entire national area. The entire industrial area north of the Loire, along with Brittany and a broad zone along the Atlantic coast, was occupied territory.

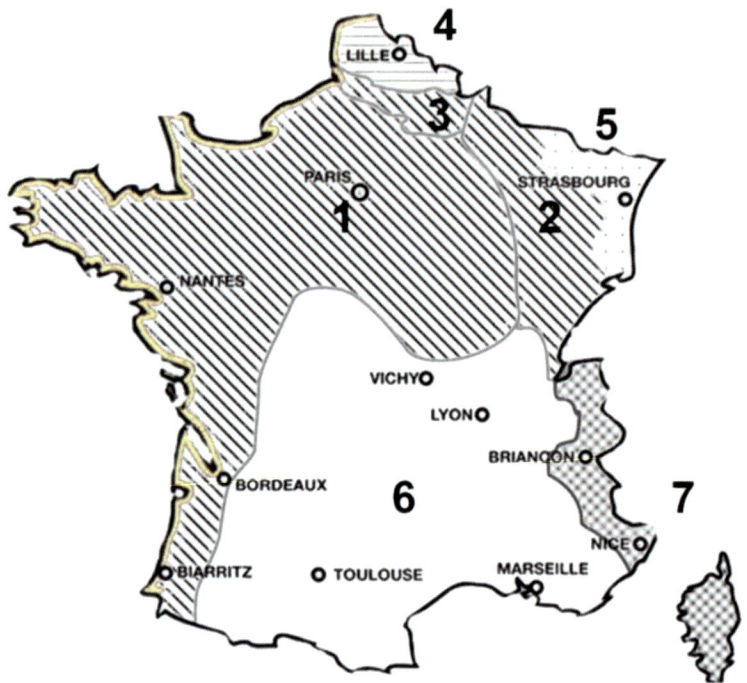

Figure 6.1. Occupied France 1940-1943. **1.** Occupied zone. **2.** Reserved zone. **3.** Prohibited Zone. **4.** Governed by German Army. **5.** Annexed by Germany. **6.** Free zone. **7.** Demilitarized zone (and Italian supervision extending to Rhone)

The Germans made a very peculiar arrangement with the Vichy state: its administration would not only rule in Vichy territory, but also govern the occupied territories. Only one zone was excluded from this arrangement: a 'closed' area running along the northern boundary of Switzerland to the Pas-de-Calais coast (2,3,4). This area was likely to be annexed in the future by Germany, and Hitler claimed even more territory, but the course of war never allowed him to definitively make up his mind.[7]

The outcome of the Second World War 'Act I' was a political body, unique in Western Europe. The French Third Republic (established in 1870 and abolished with Germany's invasion in 1940) had collapsed into political unit that claimed to be a legitimate successor state, but within an international system dominated by Germany. This differed from the response to the German occupation elsewhere. In the Netherlands, for example, the government and queen fled to England during the first days of the German invasion. The Dutch state nonetheless continued to exist as an administrative system and required officials to collaborate with the German military administration. The governing situations in conquered states varied along a continuum between outright collaboration in executing Germany's war aims to, what became a Dutch expression, 'mayor in wartime': remaining at one's post to serve the civilian population and preventing even worse outcomes. The acceptable limits of the latter became a bone of contention after the war, but continuity of a state's leadership raises even more tricky questions about legitimacy.

In contrast to the Dutch scenario, the Vichy regime recruited leaders who bore responsibility for the war efforts and state affairs in the tragic days of the German invasion and who remained at their post. There had been a changing of the guard during the critical days in June 1940 when the government was deeply divided on the question of whether to form an armistice or continue the resistance. Prime Minister Paul Reynaud had stepped down from office and President Lebrun appointed the aged Henri-Philippe Pétain, hero from the First World War. He was the man who would lead the French government into Vichy. It in no way resembled anything like a coup. The Assemblée Nationale (then the two Chambers of representatives) gave Pétain full powers to draft a constitution and establish a new government that essentially abolished representative democracy. The country was in a state of crisis and well-considered political action had become a distant ideal.

Why did the French government choose the dishonourable option of a subservient state? There were many reasons to do so. The most direct one was the loss of life at the war front the month before. It revived memories of the destruction caused by the previous war and ignited fear about the fate of

---

[7] Norman Rich, 1974, *Hitler's War Aims. II The Establishment of the New Order*, pp.197-237.

France's cultural heritage. A second reason was the very negative assessment of the political climate in the preceding Third Republic. The country had neglected its army and had morally suffered under a secular regime that had given matters of individual wealth and enjoyment priority over national strength and the family, as low birth rates revealed. This was the particular complaint that Pétain voiced in the run up to the armistice. No victory was imaginable unless French could retrace its steps, but such a process of national rebirth required time. The third argument followed from a 'realistic' geopolitical calculation. Germany already seemed to have won the war in Europe and it was only a matter of days till Britain would fall. France would do better to quickly devise a proposal for its semi-independent role in the New World Order to be ahead of its possible erasure from the map. The French overseas empire still guaranteed France a role as gatekeeper to the colonial world and the weak underbelly of the British Empire.[8]

French intellectuals evaluated the events in Europe as an earthquake that, although producing enormous human misery, was the signal of a new step in human evolution that would destroy the old nationalistic state structures and ultimately create a new and better Europe. As the socialist Claude Jamet wrote in 1942:

> Hitler upturns Europe, in a systematic way. He progresses like a natural force; he wins ground like a forest fire. And the small nations, at the sound of his pace, cross themselves superstitiously ... It is a curious world in which we live. The frontiers jump, the map moves, the world transforms before our eyes. I cannot believe that the will of one man, whoever he may be, can explain and determine such phenomena. Hitler is in no way the Supreme Ruler [*démiurge*] of these things. It is much worse: he is their minister (in the sacred sense), priest, elected prophet... I am raving and know it very well. So strong is the temptation, the prestige. From the moment that destiny is revealed, there is no morality anymore, no courage. Resistance appears to be in vain and foolish: can one resist geological events that mark the coming of a new era, even if it is plain that our species will not survive?[9]

---

[8] Michael Heffernan, 2005, 'Geography, empire and National Revolution in Vichy France'.
[9] Claude Jamet, *Carnets de déroute*, 1942. Quoted in Bernard Bruneteau, 2003, *L'Europe Nouvelle' de Hitler. Une Illusion des Intellectuels de la France de Vichy*, p. 41. Author's (GD) translation.

Finally, the Vichy option was a logical consequence of a certain conception of the state and government legitimacy. According to Pétain:

> It is impossible for the Government to emigrate, to desert, to abandon the territory of France. The duty of the Government, regardless of what happens, is to remain in the country or else it will not be recognized as such. To deprive France of her natural defenders in a period of general disarray is to deliver it to its enemy. It is to kill the soul of France and consequently to render impossible its renaissance…
> We must accept the suffering which will be imposed on the country and on her sons. The French renaissance will be the fruit of this suffering (…)
> The armistice in my opinion, is the necessary condition for the perpetuity of an eternal France. (June 13, 1940)[10]

Although this logic seemed to have appealed to the majority of the French people and officials at the time, not everyone agreed. De Gaulle, just promoted to the rank of general and still a minor figure in the French state elite, spoke these words from London in a radio broadcast on the June 18th.

> But has the last word been said? Is all hope gone? Is the defeat final? No!… For France is not alone (…) She has a vast empire behind her. She can form a bloc within the British Empire, which holds the seas and continues the struggle. She can, like England, utilize without limits the immense industry of the United States.[11]

No wonder that the Vichy government labelled de Gaulle a deserter, who would perhaps be executed if he were to enter their territory. Two conceptions of state legitimacy, one might say two *lines of flight*[12], were apparently clashing. Pétain's ideas owed much to the notion of territorial identity that had emerged during the Third Republic. The shock of the Franco-German war of 1870 and the loss of Alsace-Lorraine transfigured France's self-conception as a European vanguard of revolutionary principles – liberty, equality, fraternity – into that of a state that owed its personality to its territory, or as French geographers like Paul Vidal de la Blache (1845-1918) said: its *genres de*

---

[10] William L. Shirer, 1970, *The Collapse of the Third Republic*. London, William Heinemann / Secker & Warburg, p. 773.
[11] Ibid., p. 833.
[12] According to the meaning given to 'line of flight' by Deleuze and Guattari, 1992.

vie (ways of life).[13] Pressure groups that acclaimed the benefits of colonial life during the last decades of the 19th century did not succeed in making the 'soul of France' (in the words of Pétain) imperial like that of Britain. The principle that professes control-sovereignty as the hallmark of legitimacy is diametrically opposed to the principal of legitimacy by territorial governance. Only a government that has undivided authority can be called legitimate, even if this means that most of its governmental power cannot be implemented because the government lives in exile. The goal of Pétain and his council was obviously to preserve a close relation with the people and their territory. The Vichy state literally worshipped rural France; as Pétain remarked: 'the earth [*la terre*] cannot lie'.[14]

When the fall of Paris was imminent, the French government fled to Bordeaux, far from the turmoil of war. Here its members struggled with the future course of France and important talks took place with representatives of overseas allies, Britain and the United States. Even Churchill paid a visit, hoping to make the best of a bad job and keep the French fleet out of German hands. After the armistice was signed, the government again had to move because the entire Atlantic coast came under German control. They moved to Clermont-Ferrand, in the centre of France, were vice-president Pierre Laval had extensive business interests. It quickly became apparent that this industrial city had little room to accommodate the new administration and its civil servants. Therefore, they moved to near-by Vichy, a spa with a lot of vacant hotels and residences. It fit the spirit of the Vichy government even better than a big city ever would have.

The Vichy *state* was in many aspects a rejection of the Third *Republic*, but at the same time it was modelled after the foundations of the preceding state. The name change was significant: it was a state, no longer a republic. This reflected the contempt that many intellectuals had felt for parliamentary politics in the 1930s. The instability of the Third Republic was proverbial; it claimed more than one hundred presidents in the seventy years since its inception and more than 20 governments in the mere seven years before the war. The Vichy state, in contrast, was not going to be a representative democracy. Its president acquired both executive and legislative power, a sin against democracy, but even the Third Republic had in its last years run into

---

[13] M.-C. Robic, 1995, 'National identity in Vidal's Tableau de la Géographie de la France: From Political Geography to Human Geography'.
[14] François-Georges Dreyfus, 1990, *Histoire de Vichy*, p. 219.

this practice. The point is that although the president of the Republic theoretically had extensive powers according to the constitution of 1875, these usually were not executed.[15] The Third Republic was more liberal than required by the constitution. The strong state was and is a profound principle in French political thought, but it can be implemented in different ways. During the first decades of the Third Republic the state engaged in a fight against ideological enemies like the Church. Religious instruction in schools was forbidden and funds were allocated for greatly expanding the number of public schools (the ideal of *laïcité*). The government caused quite the public outcry at the turn of the century by closing all primary schools maintained by religious orders.[16]

These were not Pétain's ideas about the strong state. He gave the Church access to the people and their education, but denied people access to the state. Ordinary people became only participants through territorially binding networks of work and family. In this way they could 'influence' the state and practice their citizenship. The motto of the revolutionary tradition in France, 'liberty, equality, fraternity', was replaced with 'work, family, country [*patrie*]'. Children were to be educated in the country's cultural heritage by visiting pilgrimage sites, castles, and cathedrals[17] while their parents were made to repent and procreate. It looked like a step backwards to the moral government of early modern Europe (in which social problems were attributed to individual vice), but it was essentially a deeper implementation of the governance-identity state. Many of the virtues of post-war France were actually started by the Vichy state: the *périphérique* (ringway) that encircles Paris, the administrative reorganization in regions, the national police, the various policy fields of the welfare state (public health, old-age pensions) etc.[18] Far from silencing renowned intellectuals, this philosophy prompted many of the best and brightest to adapt their thinking the new focus. Even representatives of the frivolous sector, like fashion designer Coco Chanel, were swayed by Vichy's mindset.

The psychological mood of the period is critically analysed in Jean-Paul Sartre's play 'The Flies' (1943). The Flies criticizes the disciplining of

---

[15] Ibid., pp. 207-209.
[16] William L. Shirer, *1970*, pp. 51-54.
[17] John Hellman, 2001, 'Memory, history and national identity in Vichy France'
[18] Cécile Desprairies, 2013, *L'Héritage de Vichy : Ces 100 mesures toujours en vigueur*.

> **Box 6.1.** Repentance in Sartre's *The Flies*
>
> The Flies is a variant on the Orestes story in which Orestes after long wanderings returns to his homeland, the city of Argos, only to discover that it has fallen into a state of doom, symbolized in Sartre's play by a swarm of big flies. Argos has fallen to the dictator Aegisthus who, after killing its legitimate king Agamemnon, has married his wife (and Orestes' mother) Clytemnestra. The wartime audience immediately recognized these roles as symbols for the German occupier (Aegisthus) and Pétain (Clytemnestra). Like Pétain in Vichy, Aegisthus has organized a collective ritual of repentance, creating a society weighed down by 'meaculpism' on the death of their sovereign (the neglect of the state and army in the Third Republic). In order to liberate his compatriots, Orestes has to break the spell of hereditary sin and ultimately kill Aegisthus and Clytemnestra. The particular additions made by Sartre are a God (Jupiter) that conspires with the dictator (referring to the privileged status that the Church was assigned in Vichy) and a very weak Electra (the Vichy intellectuals?), Orestes sister, who hates the regime yet cannot free herself from its mental climate. When Electra tries to pit the people against the regime with provocative behaviour (with the goal to prove that no Gods would demand self-pity and lack of human enterprise), Jupiter intervenes by rolling an immense rock from the mountain. The community is again intimidated and turns against Electra and the stranger Orestes. Yet, Gods are incapable of acting against people who do not go along with common beliefs and are the captains of their own soul. Orestes cannot be stopped from killing Aegisthus and Clytemnestra. After the deed he leaves the city, freeing it from its evil (the flies).

society with a combination of religion and guilt about the past. Sartre's (existentialist) message is that people should **take their individual responsibility even if it clashes with social norms and** values. God approves of dictators (who keep people religious), but the might of these divine and human conspirators is limited when they face a self-conscious human being. It is no wonder that existentialism became the favourite philosophy, or rather the favourite 'imago', of intellectuals and young people in France immediately after the war. They plunged into a ritual of purification that was the reverse of Vichy: celebrating nonconformity, creativity and … the city. Sartre himself wrote, 'A city is a permanent creation: her shop-windows, smells, sounds and to-and-fro represent the human regime. Everything in the city is poetry in the narrow sense of the word'. Everything in the countryside, according to Pétain, was patriotism and piety. There is obviously a territorial logic in dramatic rejections of the political order. [19]

When allied forces marched into North Africa at the end of 1942, German troops occupied the 'free' Vichy zone. The last scrap of sovereignty was lost, yet the Vichy regime remained in power. Its legitimacy was further undermined by German demands regarding local labour supply and the transfer of Jews. Truth be told, the regime had to cope with an ever more aggressive resistance movement. In December 1943 Pétain resigned because he could no longer go along with further incursions on the regime's freedom to act. Laval became head of government. After the German defeat, both leaders were arrested (along with many other figures that had actively collaborated with the Germans). Laval was executed in 1945 and Pétain, after a reprieve awarded by general de Gaulle, was imprisoned on the island Ile de Yeu. He died in 1951 at the age of 96.

De-territorialization was manifest in two lines of flight: one was the escape of a sovereign France impersonated by De Gaulle, the other an escape into France as a territorial personality. The latter was the line of those who maintained that France could not be destroyed by an occupation but had to be healed in order to be vital in a new Europe; a *territorium* was to be temporarily turned into a *sanatorium*. Since the main handicap of the defeated French was sovereignty, it had to mobilize other territorial visions in order to recover. Overcoming this handicap required the vision of a supranational Europe in which the role of nations would depend on geography (economic specialization) or identity (cultural heritage). Such visions had already become popular in France between the World Wars. They had diverse sources of inspiration: the peace ideal sprung from abhorrence of the First Wold War slaughter, the general failure of the Third Republic, the failure of state control in general vis-à-vis the economic crisis and international capital, rejection of the two 'materialistic' models offered by the East (communism in the Soviet Union) and the West (American capitalism), the revival of Catholicism and its old imperial ideal, the slow transformation of democracy into technocracy, etc. Vichy elevated such visions to relevance but could not openly promote ideas

---

[19] Michel Winock, 2001, 'Sartre: l'effet de modernité'. 2 (p. 204). David Drake, 2002, *Intellectuals and Politics in Post-War France*, p. 24 (n. 50).

about a future in which the role of Germany was reduced to that of a member state in a federal Europe. Yet one could covertly speculate about the role of France in a nebulous future superstructure in which sovereignty in the absolute sense would be eradicated. This usually amounted to three aspirations: France as the tradition-carrier of Roman and Greek civilization, France as the vanguard of a revitalized spirit of Christianity, or lastly, France as international mediator.[20]

In one place, the Vichy-established school for officials and youth leaders in Uriage (near Grenoble), discussions advanced quite far in the direction of a federal Europe or what would become the European Community after the war. One of the speakers in 1942, the young professor of International Law Paul Reuter (1911-1990), suggested that the individual states would be assimilated into a European organization with a system of double representation: one according to importance and the other in terms of equality (the federal principle). Among the collective institutions he mentioned were a federal army, guaranteed rights for nations and individuals, abolishment of customs, a central bank, and European public services such as communications, justice, private law and a budget office.[21] As Julian Jackson remarks: 'Many of those who had worked at Uriage enjoyed greater intellectual influence, especially in journalism and publishing, after the Liberation than they had ever done under Vichy'.[22]

When the war ended with an unconditional surrender of Germany, such thoughts were pushed into the background. The division of Germany into occupation zones was underway and compensation for war damages was at the forefront of the European mind. Particularly the industrial heart of Germany, the Ruhr area, was to be internationalized, while its industrial establishments would be dismantled and the production capacity transferred to allied countries. At the same time, the deterioration of American-Soviet relations made the US more sympathetic toward the idea of German economic recovery, thereby creating a stronger Western bulwark against communism. France, in need of Germany's coal, did not agree. The idea of Marshall Aid was proposed, partially, to eliminate French worries about economic recovery, but could not sufficiently reassure them.[23]

---

[20] Bruneteau 2003, pp. 179-189.
[21] Ibid., p. 201.
[22] Julian Jackson, 2001, *France: The dark years 1940-1944*, p. 599.
[23] Alan Milward, 1992, *The European Rescue of the Nation State*.

Chapter 6. DYING STATES

In 1948, the Soviet Union denied the other allies access to West Berlin and withdrew from Allied Central Control, and the whole idea of divided control over German territory by the Western allies fell to pieces. In this situation of geopolitical emergency, the French international representative Jean Monnet first launched the idea of a 'federation of the West', an International Authority of six powers (US, Britain, France and the Benelux countries) that would oversee the Ruhr area production of coal and steel. Because the International Authority did not curb tensions stemming from French and German desire to resume their roles as autonomous players (basically supported by the US), Jean Monnet and the French foreign minister Robert Schuman finally proposed a European coal and steel community. This coalition was the first implementation of a federal Europe as conceived by certain intellectuals in the interwar period (and more existentially voiced during the years of occupation). According to Michel Loriaux, it was an *'ontopological'* shift from an 'Ossian' (nationalist) to a 'Carolingian' discourse, in which international relations and geopolitics were transformed into the relations *within* a transnational community (empire): the deconstruction of the Rhineland as frontier.[24]

To succeed, Monnet and Schuman's proposals needed the backing of the post-war French government and French public opinion One may only wonder whether the first steps to European integration were either the result of contingent geopolitical pressures and national emergencies[25] or the adaptation to an inescapable new territorial order. The Americans (and the Soviets) were initially not inclined to give the Europeans a united voice, as this would only complicate their dealings with occupied Germany. However, after the drifting apart of the two great war victors, a unified Western Europe emerged as a strategic asset for the US. Nowhere else (for example in Japan and Pacific Asia) has such a development been replicated among previously independent states. A long history of intra-European dealings between states (the European 'concert'), inspired by mutual feelings of fear but also based on mutual cultural recognition, was of course a facilitating condition that did not exist elsewhere. The confrontation with new powerful global actors outside the European arena also made European identity more tangible. **In this sense, supra-nationalism was an important internal factor in European**

---

[24] Michel Loriaux, 2008, *European Union and the deconstruction of the Rhineland frontier*. The term 'ontopology' is taken from the French philosopher Jacques Derrida and means the essence (being) of something as derived from its 'topos' or situation.
[25] As strongly argued by Alan Milward, 1992.

integration; such sentiments did not exist in other parts of the world where military occupation extended across several countries. It in any case demonstrates that resurrection did call up a new state orientation in terms of control, identity, or closure.

It is not surprising that general De Gaulle, the man whose line of flight so strongly deviated from Vichy, who represented French resistance from the outside, was such a staunch defender of the old Europe of independent nations. He retained and advocated this viewpoint in his later presidential dealings with the European administration and the (European) council of state leaders. His 'portable' control-sovereignty principle aimed to make France independent from any global power (whether the US or the Soviet Union) and considered the members of the European community as an alliance reinforcing France's position. That did not alter the fact that the cultural-territorial conception of France, with its roots in Vichy and visions of Europe as an organic community, was also vivid in French political thinking. Vichy had cut off the stranglehold of the state (the Third Republic) on the cultural sphere, and religion in particular. It instead emphasized religion as penance or political purification and offered a 'salvationist teleology' that struck a sympathetic note among other European leaders (many of them Roman Catholics) who were likewise eager to get rid of the demons of the past. Since this moral revitalization and religious persuasion distinguished Europe from the great powers, it resounded with de Gaulle's aim of making France neutral with (or without) European support. Ultimately, practical constraints (like balancing the relationships with Germany) forced de Gaulle to reconcile himself to at least some European federal principles.

There was a third line of flight that was less territorial than those pursued by Vichy and De Gaulle: Sartre's transnationalism with a communistic and urban persuasion. Whereas the first line became compromised by the crimes of the Stalinist regime, urban transnationalism might be a third way out of the continuing political quandary of competing nationalism and European integration. It draws attention not only to the history of Europe as a civilization (which began in Roman times as string of towns along the Rhine axis), but also on the cross-border networks of cities that essentially constitute the engine of economic growth. It is local and global at the same time, or '*glocal*' as the current fashion has it.

The break-up of an Empire: The post-Soviet state

In the decades between 1960 and 1990, foreign observers and Russian dissidents alike concluded that from a rational point of view the Soviet Union could not go on any longer in the same way, but nevertheless had to acknowledge that as a matter of fact it continued to do so. There was widespread agreement on the inefficiency of the Soviet economy and the overwhelming costs of a system that had to repress and spy on millions of people. And yet there seemed to be an invisible and unlimited capacity in Soviet society to endure such practices. It induced Sovietologists well into the 1980s to expect no structural change in the near future. But suddenly in 1989, during the time span between conceiving such ideas and the moment they appeared in print, there was a complete transformation of the Soviet Union– to the embarrassment of many an author.

When Soviet dissident and historian Andrej Amalrik in 1969 published his essay *'Will the Soviet Union Survive until 1984?',* it was hailed in the West as a correct analysis of what was wrong in Soviet society, but not as a very likely prediction. It turned out to be rather the reverse. Amalrik realized that a break-up of the Soviet Union would be only conceivable if initiated by some shocking event. His future scenario assigned this role to war with China. Not a bad idea in terms of state collapse theories that stress the role of external events. It was a timely idea as well in view of the rising tensions between the two champions of international socialism. Ruling out the possibility of nuclear war, Amalrik had to think up some way in which the Chinese, apart from their sheer number, could be a military threat. This, according Amalrik, was the tactics of guerrilla war, which the Russians themselves had applied with success against invaders from Napoleon to Hitler. Here, the bright young historian was clearly at fault, since a typically defensive strategy used by the Chinese could hardly be threatening to the Soviet Union. Moreover, he did not seem to have studied the history of failed Chinese attempts at expansion. In another prediction Amalrik was on the right track. When the Soviet armies were entangled in clashes at the Chinese border, the satellite states of Eastern Europe would seize the opportunity to de-Sovietize (reunifying Germany) and subsequently national republics would break away from the Union: first the Baltic republics, the Caucasus and Ukraine, then also in Central Asia and the Volga area. This is a stunningly accurate description of the tide of nationalist mobilization that would flood the Soviet Union between 1988 and 1991. But little could Amalrik foresee that the initial shock would be an internal event: Gorbachev's decision to modernize the Soviet economy and society.

How important was the 'Gorbachev factor'? Would the Soviet Union still be with us if Gorbachev had not risen to power? These questions are difficult to answer; one can only say that the mismatch between the modes of production in the West and the Soviet Union was quickly growing in the 1980s. Western countries experienced a post-industrial revolution epitomized by the widespread use of personal computers and the burgeoning of small firms in which workers were also shareholders or owners.[26] Furthermore, new threats like environmental pollution commended cooperation between countries (rather than prompting the erection of iron curtains). And finally: president Reagan's boosting of the arms race made catching up with the United States appear impossible. Anyhow, Gorbachev saw the urgent need for modernizing the Soviet economy (*perestroika*). To conquer the resistance, which was propelled by established interests, he soon felt that it was also necessary to introduce another strategy. By stimulating a more open climate of discussion (*glasnost*) he hoped that fixation on self-interest would be exposed and disgraced. It was naïve of the Secretary General to ignore the other processes of change that could be unleashed by *glasnost*, like human rights and local autonomy, but he was perhaps right in supposing that it was the only chance to revitalize the ossified economic system.

In the years after the dissolution of the Soviet Union (SU) many commentators claimed that it was 'inevitable' in view of the highly explosive building bricks of the union: the federal republics with their national identities. However, as Mark Beissinger has argued, this was a false argument because a similar scrutinizing of the structural conditions of the SU ten years earlier on the contrary induced political analysts to call a break-up 'impossible'.[27] Wisdom of hindsight does not lead to reliable generalizations. Yet, it cannot be denied that national mobilization was instrumental in bringing the Soviet Union down. As in other more constructive cases of national mobilization, the central question is: why does it happen at certain moments and fail to emerge at other moments? Beissinger's analysis makes clear that the secessionist drive in the SU was possible because it operated as a 'transnational tidal force'. This involved the spatial diffusion of a 'secessionist frame' making nationalist de-

---

[26] Manuel Castells, 1998, *End of Millennium (The Information Age: Economy, Society and Culture III)*.
[27] Mark R. Beissinger, 2002, *Nationalist Mobilization and the Collapse of the Soviet State*.

mands much stronger in places where one would have expected little activism because of the small group numbers, their low degree of urbanization or their near-complete assimilation into the Russian culture and language. As in other cases of state crisis and collapse, we cannot say that ethnic diversity is itself sufficient to cause a state crisis. In fact, ethnic divisions are created during the crisis and conflict.[28]

What made the secessionist mindset so contagious in 1988-1991? There seem to be two main causes. First, the nature of particular Union Republics, like the Baltic republics, where the secessionist frame was pioneered and where a memory of having been a separate nation state was still alive.[29] Second, the special nature of the dynamics introduced by Gorbachev's *glasnost* policy, which resulted in state repression (of protests and criticism) was no longer functioning. On the contrary, even the mildest attempts by the central powers to contain social unrest inflamed protests and hastened the designing of plans for autonomy.

Of course, the special politico-geographical nature of the Soviet Union was a significant factor in this all. There has been much speculation about why Lenin and Stalin decided to institutionalize a feature (nationality politics) that was anathema to communist ideals (it rather belonged to the false distinctions of capitalism). One reason was the spirit of the times: the ideals of freedom and self-determination that permeated Europe at the end of the First World War legitimized the Bolsheviks and other European national liberation movements. In an international socialist world, nationalistic distinctions were supposed to automatically dissolve in a new world order in which boundaries would fade away. But the early communist state was weak, and its leaders jumped at any opportunity to create bonds of solidarity, even if it meant recognition of ethnic politics. One may ask then why this principle was not abandoned when the Soviet state grew stronger. In certain cases, promises made to allies in the revolutionary struggle, like the Volga Tatars, were indeed reversed. In a 1918 issue of the Pravda daily, the aspirational plan for a Tatar-Bashkir Soviet Republic encompassing most Tatar-speaking people was still mentioned, but two years later the result was an autonomous republic that

---

[28] Caty Clément, 2004, *State Collapse: A Common Causal Pattern? A Comparative Analysis of Lebanon, Somalia and the Former-Yugoslavia*, pp. 49-55.

[29] The cover of Beissinger's book shows a picture of a rally held in August 1989 in Tallinn, in which a banner is carried saying 'Estonia never joined USSR!!' (i.e. it was forcefully incorporated after the Molentov-Ribbentrop pact).

left 75 percent of the Tatar-speaking population out.[30] This outcome was typical of geopolitical considerations behind the Soviet nationalities policy. On the surface they openly recognised national identities, but then drew boundaries behind closed doors in such a way that the political (territorial) strength of a national group was effectively diminished (comparable with the practice of 'gerrymandering' in the delimitation of voting districts). There was also a more 'sincere' motive behind the continuation of the nationalities policy: the fact that cultivation of a national language was necessary for education. Education was necessary to create a ruling class that would be committed to the cause of communism.[31] In some cases the merit of a nationalist policy is obvious, for example where a group had no native writing tradition and something like a script first had to be invented. Anyhow, the idea was that through the process of nativization (*korenizatsiya*) and its subsequent intellectual development, the nations would converge (*sblizheniye*) and finally amalgamate (*sliyaniye*).

The structure that arose from these considerations was asymmetrical. It involved different levels of autonomy: Union republic (15), autonomous republic (20), autonomous '*oblast*' (province, 8) and autonomous '*okrug*' (district, 10). Union republics like the Baltic republics, Ukraine, Georgia, and Azerbaijan were a gesture to nations with a comparatively high level of modernization and/or memories of a former existence, however short, as independent states. They were allotted the highest degree of autonomy, even the constitutional right of secession (on which they eventually appealed when disruptions of the political order intensified at the end of the 1980s). The most problematic entities were the autonomous republics like Chechnya, Abkhazia or Tatarstan. After the dissolution of the SU, these former autonomous republics continued their existence within the new states (Russia and Georgia) but without the former protection of a supreme authority (the Soviet state) and were consequently entangled in conflicts with the newly independent states to which they formally belong.

What we cannot see in the death throes of the Soviet Union is the opening of a 'prison of nations'. One should not overlook the early Bolsheviks' contribution to the successful emergence and maintained presence of 'nations' and their territorial habitus, and, secondly, how much their emergence in the 1990s again was the product of a unique mobilization process.

---

[30] Azade-Ayşe Rorlich, 1986, *The Volga Tatars. A Profile in National Resilience*, p. 136.
[31] Robert J. Kaiser, 1994, *The geography of nationalism in Russia and the USSR*.

Territorial independence was *not* the first thought that occurred to the citizens of the Baltic States, Ukraine or Georgia. That idea only hesitatingly emerged after cycles of (weak) oppressions and mobilization culminating in the gradual loss of legitimacy of the central government and its military apparatus. The break-up cannot simply be equated with the awakening of dormant primordial '*ethnies*'. In the end, Russian residents of the Baltic republics and Ukraine favoured independence just as much as the ethnic nationals.

In the demise of the Soviet Union, we observe one of the clearest examples of how de-territorialization operates by using the very principles and institutions that previously constituted the backbone of integration or consent (*governance*). Apart from the federal system itself, one such principle was the 'institutionalized' mode of repression in the Soviet Union. For more than six decades, the SU succeeded in preventing violent action against the multitudes. Usually, the combination of the surgical arrest of a few key leaders, the harassment of writers and intellectuals, the seizing of banners and leaflets and the presence of an overwhelming police force was sufficient to prevent disorder at mass meetings. But glasnost had defused such strategies. For one thing, the number of meetings had grown to such an extent that control became difficult. The Gorbachev regime certainly attempted regulation of mass events. The Soviet constitution guaranteed its citizens freedom of assembly in private and public places, but the law required that organizers of mass meetings had to ask permission. But since permission was never given in the past, the implementation of the law under the influence of *glasnost* meant that permission was rarely sought.[32] Soviet citizens simply had no experience with democratic regulation by law. There was one predominant guiding principle: the incompatibility of absolute freedom and repression. In the first nine months of 1989, 724 mass events took place in Ukraine but only 47 percent were authorized.[33] Attempts to enforce the law only made things worse because it implied that the regime was working to reverse liberalization.

On April 9 1989, a violent crackdown on a demonstration in Tbilisi sent a message all over the Soviet Union. However, the crackdown backfired and in the end intimidated not so much potential protesters but rather the military and their leadership. The motive for repression was understandable: the Caucasus threatened to become a hotbed of ethnic violence. The rumour that force was going to be used in Tbilisi in order to repress the demonstra-

---

[32] Mark R.Beissinger, *2002*, pp. 334-342.
[33] Ibid., p. 341.

tion had spread in advance, but this only drew more protesters. Events escalated and evolved into a clash between the military and the demonstrators in which twenty people were killed and many more injured. The end result was that Tbilisi became a symbol of illegitimate action by the Soviet government. At a counterdemonstration to the traditional commemoration of the Bolshevik victory in Chisinau, members of the Moldavian Popular Front called upon 'the army of Kabul and Tbilisi…[to]… repent and restructure yourself'.[34] After April 1989, the deployment of troops was significantly constrained, both because of hesitations among the Soviet leadership and resistance of the military to perform police tasks, which they deemed 'humiliating'. After Tbilisi, even mass violence and destruction by demonstrators would go without punishment.

As remarked above, we should not interpret the institutional policies that characterized the founding of the Soviet Union as in some way recognizing the principle of the nation state. *Sliyaniye* aim was not to weld all Soviet peoples together into one cultural nation, either a new one or, notwithstanding the propagation of Russian as a lingua franca, one based on Russian 'higher' culture. As Rogers Brubaker aptly remarked: 'No other state has gone so far in sponsoring, codifying, institutionalizing, even (in some cases) inventing nationhood and nationality on the sub-state level, while at the same time doing nothing to institutionalize them on the level of the state as a whole'.[35] Cultural identities did not play a role in Soviet ideology; consequently, the 'national' republics were never intended to provide a territory ripe for the nationalizing (culturally homogenizing) impulse by their governments.

The creation of national republics was a highly ambiguous act because it went along with another, personal, way of defining national identity. Nationality was an attribute of every citizen of the SU, affirmed in their passports, irrespective of the place of residence of the person concerned. This means that Russians living in Kazakhstan or Ukraine or Georgia would keep their Russian nationality. Such personal nationality did not give specific political rights. The republic, defined in national terms (based on its 'titular nationality'), gave leeway to an ethnic administrative class that still had to share their authority with representatives (often Russian) of the central communist

---

[34] Ibid, p. 354.
[35] Rogers Brubaker, 1994, 'Nationhood and the national question in the Soviet Union and Post-Soviet Eurasia: an institutionalist account', p. 52.

party. Autonomous Republics, which were encompassed by the territory of a Union Republic, further checked the freedom to give free rein to nationalist policies by a Union Republic's government. Some authors see a hidden purpose in the geographical delimitation of the Republics, for example in the way the Armenian exclave Nagorno-Karabakh (in Azerbaijan) was cut off from the Armenian motherland, or the ignorance of the distribution of the Tatar-speaking population in separating Tatarstan and Bashkiristan. 'Divide and rule' was applied to both internal nations and nations outside the Soviet Union. Examples of the latter are the creation of the Union Republic of Azerbaijan, possibly intended to pit Iranian Azeris against Iran, and the Union Republic of Moldavia being pitted against the Moldavian minority in Romania.[36] That the outcome on the map can be explained by a geopolitical divide-and-rule strategy does not mean that there is documentary evidence to prove such intentions. Whatever the background of the creation of these units, they fit the general conclusions that **the ethno-political geography of the Soviet Union was not intended to nurture local autonomy but to integrate the state. The irony was that it offered a most useful tool for disintegration as well.**

The Soviet Union was a peculiar blend of state and empire. On the closure and governance dimensions it met the criteria of the high-modern state, but on identity it rather looked like an empire: an assembly of different ethnic groups which were not subjected to cultural assimilation. In a way this quality of empire had itself lodged into the Soviet citizen's identity as pride in being a member of a worldplayer state (occupying a geographical pivot in Mackinder's terms) and on having surmounted the colonial predicament and 'racism' that had disrupted the rest of the world, and particularly 'the West'. But this feeling became much more articulated in the politics of nostalgia that followed the disintegration of the Soviet state (see about Aleksandr Dugin below).

Re-territorializations proceeding from the Soviet collapse
When disappointment in the economic performance of the federal government (or the Soviet communist party) coincided with the transition to a multiparty system, the scene was set for national groups to articulate their own interests. The failed communist coup in August 1991 provided the definitive

---

[36] John O'Loughlin, Vladimir Kolossov and Andrei Tchepalyga, 1998, *National construction, territorial separatism and post-Soviet Geopolitics: the example of the Transdniester Moldovan Republic*.

first crack in the crumbling of the Soviet Union and also gave the opportunity for local communist executives to reframe themselves as national-communists. However, the wish of the Russian Federation's president Yeltsin to pursue the transition to a market economy and political turmoil in the Soviet Union cleared the way for proclaiming an independent Russian Federation no longer based on communist principles. The Soviet Union became a Commonwealth of Independent States in which Russia retained an essential stake both in terms of territorial size and distribution of people with Russian identity. The maladaptation of its productive system to a market society caused many years of economic troubles in the Russian Federation

The other seceded states from the former Soviet Union have struggled in different degrees with forces that undermine their state integrity (transnational crime, disloyal ethnic groups) or their ability to pursue independent international politics. Georgia is a case where the strength of the national movement for independence was in striking contrast to the ability to forge an integrated state. Although significant, this was not merely the result of the presence of two Autonomous Republics (Abkhazia and Adjara) and one Autonomous Oblast (South-Ossetia) on Georgian territory which were resistant to Georgian nationalism.[37] It was rather the structure of Georgian society itself that resisted the development of the civil society that is a necessary condition for political (and territorial) integration.

Georgian society was predominantly a rural society in which the family counts first and leaders rely on a clientelistic structure rather than well-organized political parties. Such structure was quite compatible with the communist one-party system as it prevailed until 1989, since the Communist Party also offered members privileges and exclusive access to goods or jobs.[38] While this may have helped Georgia to territorially integrate into the Soviet Union, it could not reproduce the same result for an independent Georgia. It is not surprising that when the Communist Party lost its legitimacy in the wake of the Tbilisi massacre, a situation devoid of law and order developed in which paramilitary groups struggled for power. During the Gamsakhurdia era (1991-1992), leadership was based on symbolic nationalist gestures rather than effective and representative government; government forces were 'little

---

[37] Only in Adjara where the population considers itself Georgian no violent conflict burst out.
[38] Jonathan Wheatley, 2005, *Georgia from National Awakening to Rose Revolution. Delayed Transition in the Former Soviet Union,* p. 35.

more than another militia'.[39] The leaders of the coup that overthrew Gamsakhurdia in 1992 invited the former First Secretary of the Soviet Union, the Georgian Eduard Shevardnadze, to become the new head of state. He only succeeded in repressing the militias and the defeated former president's supporters by calling on Russia. This meant that Georgia had to join the CIS (Commonwealth of Independent States), a decision that, from the perspective of many Georgians, was a critical deviation away from the road to independence.

Saakashvili's 'rose revolution' (2004) ended a regime that, although did not suppress freedom of opinion, continued to rely on obscure networks of influence, including the old communist 'nomenklatura'. The new regime vowed that it would reinvigorate civil society and more emphatically declare its alliance with the West, and particularly the United States. As figures 5.2-5.4 illustrate, the Saakashvili government has performed a miracle by swiftly and energetically attacking the ingrained 'cultural predilection' for corruption, even against prevailing public opinion about the necessity of gradual change.[40]

However, one may fear that economic growth stagnation (or negative growth) may endanger political stability and resuscitate old habits. Georgia's attempted restoration of the ancient 'Great Silk Road', the route that in medieval times connected the West with the Far East, while it simultaneously ignored Russia, did little to solve immediate problems. As Russians were keen to remark,, Georgia's global aims and cultural and political alliance with the West overlooked the fact that the development of business 'inevitably ties Georgia to Russia'.[41] It remains to be seen whether the similar Chinese revival of the Silk Road in the so-called *Belt and Road Initiative* (an intended connection of China over land and sea to Eurasia/Europe) will help to advance Georgia's independence. In that case, the early revival of this vision in Georgia would have been clairvoyant.[42]

Economic integration with the former Soviet empire was decidedly more difficult to cast off than political and cultural heritage. Georgia to this day almost exclusively depends on Russia for mineral and energy resources;

---

[39] Ibid., p. 59.
[40] Johan Engvall, 2012, *Against the Grain. How Georgia fought Corruption and what it Means*.
[41] Felix Stanevsky, Russia's ambassador in Georgia, on 26 January 2004. *The Current Digest of the post-Soviet Press* 56 (2004) 4.
[42] Joseph Larsen, 2017, *Georgia-China Relations: The Geopolitics of the Belt and Road*. Stanevsky (2004) mentions the revival of the Silk Road ideal as something that already started under the Gamsakhurdia regime (see preceding note).

it immediately felt the shock of having to pay international market prices for gas and petrol after its independence in 1992. The metropolitan city of Tbilisi was not organically linked with the Georgian economic space but rather with the all-Union industrial network. During the Second World War, military factories were evacuated from European Russia and the Ukraine for security reasons and subsequently established in Tbilisi.[43] The military-industrial complex lost its economic meaning for Georgia after secession. The Saakashvili government withdrawal from CIS after Russian intervention in Abkhazia and South-Ossetia (2008) even further reduced links with the region.

Similar problems have haunted other former Soviet Republics as well. In the Baltic States, Estonia and Latvia in particular, the languishing of former Soviet industries hit Russian minorities hard. On the eve of the break-up of the Soviet Union in 1989, ethnic Russians made up 30.3 percent of the population in Estonia, 34 percent in Latvia and 9.4 percent in Lithuania. Because the Baltic States had been independent before 1940, they considered themselves neither new nor successor states of the Soviet Union. Consequently, their Russian (and other) minorities were defined as foreigners. Apart from their economic plight, Russians faced the threat of being excluded from citizenship in the new states because of the narrow ethnic citizenship definitions, not to mention measures that would deprive them of cultural rights like using their own language in public communications (advertisements and announcements). Criticism by the UN Committee on Human Rights, the High Commissioner for Human Rights of the Organization for Security and Cooperation in Europe and the EU prompted Estonian and Latvian governments to soften naturalization criteria (like years of residence or birth in the country) and language requirements (such as the language proficiency of candidates for parliament). In the end, the continuing pressure toward uniformity of language in public communications, for example the law that 75 percent of all television broadcasts should be in the Latvian language, may have been counterproductive. The proliferation of cable television (which was not subject to this law) enabled Russian speakers to pull in every major channel from Russia.[44] Notwithstanding their high support for secession and entry into the EU, such conditions have allegedly helped spread a pro-Moscow view among Russian speakers.

---

[43] Revaz Gachechiladze, 1995, *The New Georgia: Space, Society, Politics*, pp. 156-160.
[44] The Baltic Times 13-6-2003 (www.baltictimes.com), in Ian Jeffries, 2004, *The Countries of the Former Soviet Union at the Turn of the Twenty-first Century*.

Partly because of the complicated identity politics in the first decade of their independence, Estonia and Latvia entered the EU in 2004 with substantial stateless minorities. Authorities expected that the disappearance of this anomaly would take at least another decade, but the EU's tolerance for a situation like this in a candidate member state was in contradiction with its principles.[45] In countries that did not have the clear promise of EU membership, like Ukraine and Moldova, similar tensions between the titular nation and Russophone population (or west and east oriented territorial groups) have proven difficult to eliminate and have prevented further EU involvement. The split of Moldova into two separate parts, with a regime established in Chisinau (Moldova) and another (the Transdniester Republic) in Tiraspol, is one of the more notorious cases. The Tiraspol regime has been anxious to rely on Russian support, whereas the Chisinau regime is divided by factions that advocate amalgamation with Romania or incorporation in the EU.

> **Box 6.2. *Black hole*** (New York Times 2002)[46]
>
> A breakaway Russian-speaking province of Romanian-speaking Moldova shaped during a decade of self-rule by a curious blend of Soviet political theory and practical business ethics, the 'republic' is a ghost nation – strong enough to resist retaking by Moldova, but too weak to win statehood. It shows up on maps, but only the ones printed here, as a Rhode Island-size speck of land wedged between Moldova and Ukraine. Otherwise it exists largely in the minds of people (…)
>
> Oazu Nantoi, the program director of the Institute for Public Policy in the Moldovan capital, Chisinau, called it 'a zone uncontrolled by any state'. Symbolically speaking, Mr. Nantoi said, the region is a 'black hole.' (…)
>
> Consider weapons: with but 7,500 men in uniform, it maintains a small-arms factory for what it calls self-defense. Its products have persistently been linked to conflicts in the Balkans, Chechnya and Africa…
>
> More powerful armaments, from rocket launchers to armoured vehicles, are said to have been smuggled out of a 40,000-ton complex of weapons dumps – Europe's largest – on a Russian base here. The republic

---

[45] Eiki Berg and Wim van Meurs, 2002, 'Borders and orders in Europe: limits of nation- and state-building in Estonia, Macedonia and Moldova'. Vladimir Kolossov and John O'Loughlin, 1998, 'Pseudo-States as harbingers of a new geopolitics: the example of the Trans-Dniester Moldovan Republic'.
[46] Michael Wines, 'Trans-Dniester "Nation" resents shady reputation.' *The New York Times* March 5, 2002, p. 3.

claims a share of the aging munitions as its own, blocking international attempts to destroy stockpiles. Or look at trade: after the Trans-Dniester Republic and Moldova briefly set up a joint customs operation, 1998 figures uncovered by Mr. Nantoi showed that Trans-Dniester, with but one-sixth of Moldova's population, imported 6,000 times as many cigarettes as the rest of the country (…)

Mark Galeotti, an expert on Russian and Eurasian organized crime at Keele University in England, said that the Trans-Dniester Republic maintained an uneasy peace between five to seven international criminal gangs with varying holds on power. The lure of easy money has not only deeply corrupted this enclave, he said, but has penetrated both Moldova and Ukraine. Ukraine has resisted efforts to assert stronger border controls over the region.

The victory of the Communist Party in the 2001 elections moved Moldovan policy closer toward an alliance with Russia and Belarus. In domestic politics it started attempts to rehabilitate the Russian language in education. Moldova clearly suffers from a contradiction about its place in the world (both literally and figuratively) as it is located on the cutting line of the West-European and Russian spheres of influence. Its secession could not solve such problems, not in the least because it was unable to fall back on an earlier period of national independence or ethnic distinctiveness. The fragments of Box 6.2 taken from a newspaper article published in 2002, nicely list the features considered most characteristic of unrecognized-states: smuggling, weak internal order and danger to international order (terrorism).

The collapse of the Soviet Union has produced several of these unrecognized or *de facto* states: apart from Transdniester in Moldova there are Abkhazia and South-Ossetia in Georgia and Nagorno-Karabakh in Azerbaijan. In one way or another, unrecognized states desperately need some backing from the world outside, be it (somewhat paradoxically) the UN, a superpower, an ethnically related state (like Armenia in the case of Nagorno-Karabakh), or a diaspora.[47] Russia has openly defended the interest of secessionists in Abkhazia, Trans-Dniester, Nagorno-Karabakh, and South-Ossetia and supported the controversial change in course toward Russia in Ukraine under president Yanukovych (2013). Finally, it even militarily intervened in Georgia (2008) and Ukraine (2014), although in a disguised way by supporting

---

[47] See www.armeniadiaspora.com, www.somalilandforum.com

the military actions of rebels in the neighboring republics. All these moves create a new sphere of influence and dominance in the Russian 'Near-Abroad': one that is elicited by local national tensions for which foundations were laid in communist territorial politics during the early days of the Soviet Union. Long before Russia's intervention in Georgia, Abkhazia's president Sergey Bagapsh already pointed out that Abkhazians were tied to Russia by 'an umbilical cord', evoking the mother-child relation.[48] In such cases support of friends and ethnic kin goes hand in hand with the restoration of Greater Russia. However, one should not categorize this process too quickly as another example of immutable geopolitical laws, at least not without considering the specific influence of territorial shock.

The territorial shock caused by the dissolution of the Soviet Union was not initiated solely by the loss of territory, but also by the territory's exposure to globalization and the encroachment of other security systems (NATO) on the Russian core area. Russia experienced the turning away – to the West – of Ukraine, which after all is Russia's historic cultural heartland, as a downright attack on its national identity. It is not surprising that spiritual reactions seized on globalization as the main threat. The experience of awakening in 'another country' that has lost its values, where the only ideal is becoming a 'banker, manager or escort girl', resounds clearly from the interviews that Svetlana Alexievich conducted with ordinary Soviet citizens in the two decades after 1990.[49] While these interviews cannot claim any representativeness, the repeated complaint that one had spent a life 'building an empire... sold for a five kopeck-piece' can be easily seen as a call for revival (one that may appeal to wider circles and generations).[50] Such revival would not so much evoke the Stalinist era but rather the image of a strong state that solves the basic problems of life and relieves the individual of difficult existential choices, such as those presented by globalization. It is a double-edged sword: a positive espousal of identity (which provides instant values) combined with support for a political system that boasts prestige in the world. It is Russia's moral way of overcoming the shock of globalization, and one that Putin skillfully exploits.

---

[48] *Abkhazia's independence needs recognition before it can join Russia – Bagapsh*. Radio Mayak, Moscow 26 January 2005 (BBC Monitoring International Reports).
[49] Svetlana Alexievich, 2016, *Secondhand Time: The Last of the Soviets*.
[50] Alexievich's interviewees do not represent the generation born after 1970.

The ethnic theme in Russia's political philosophy should not be equated with traditional state-nationalism. It is rather steeped in ideas of the Russian émigré(!) – philosophers (writing in the 1920s) about Russia's Eurasian identity and their more recent followers, Lev Gumilev and Aleksandr Dugin. Their work is significant because it provides a spiritual reservoir that a leader like Putin, who is usually driven by power politics or personal interest, sometimes taps. This may be to strengthen domestic belief in the national cause or to win influence among right-wing social critics in the West. It never comes to a complete embrace of such figures, Putin's approach is too cautious for that, but it helps to endow Russian politics with a deeper meaning. In Gumilev's vision (which has indeed much affinity with the political alt-right in the West), humanity is divided into ethnic groups that have unique spiritual potency and derive their strength from a specific natural environment.[51] They are almost defined in a biological way which prescribes interaction with a specific territory (not in a political sense) and staying strictly endogamous. Since there are no universal values, all groups are equivalent. This is in contrast to the Nazi *Blut-und-Boden* theories of the past and a foundation for Dugin's attack on 'the West' and its version of globalization, which he regards as racist because it propagates universal values. The separation of the political and the ethnic is of course also convenient in declaring the continuity of a multi-ethnic Russian Empire which is based on respect for internal cultural differences but also enables a common (Eurasianist) orientation in global affairs. Using violence towards former Soviet states in the 'near-abroad' is apparently reconcilable with this vision, as the following event suggests.

In June 2008, Dugin was photographed with a Kalashnikov in front of a tank of the South Ossetian rebel army. It silently depicted the catchphrase 'Tanks to Tbilisi', which he introduced in the media. Two months later the Russian intervention in Georgia was a fact.[52] Yet in 2014, Dugin's plea for a Russian intervention in Ukraine (again calling for violence) caused him to lose his position as chair in philosophy at Moscow's Lomonossov University. This incident raises doubt about the assumption that bellicose thought will automatically find fertile ground in Russian policy circles or public opinion. Conversely, one must also acknowledge that Dugin's thought finds root in reasonable questions on the dominant political philosophy (liberalism) be-

---

[51] Mark Bassin, 2015, 'Lev Gumilev and the European New Right'.
[52] Anton Shekhovtsov, 2009, 'Aleksandr Dugin's Neo-Eurasianism: the New Right *à la Russe*'.

hind global change. All the same, in Dugin's writings there is an undeniable reminiscence of early 19th century German (nationalist) thought and of a Russian tradition most clearly embodied by the work of Alexander Herzen, who criticized the idea of progress and believed that Russia's strength is ultimately in the peasant tradition. A more recent German philosopher in this vein, Martin Heidegger, may be accurately labeled as Dugin's more explicit philosophical guide.[53]

The criticism of liberalism as a political philosophy that pursues the greatest measure of individual freedom focuses on a few interconnected issues. First, its lack of direction or vision of the future (clenched in the slogan 'the end of history'). Any world is possible in liberalism, even a virtual one or one dominated by robots. Second, the inorganic character of the ideal of growth/progress. With examples from anthropology and system analysis, Dugin emphasizes that linear growth in machines and in closed ethnic groups always requires a counterbalancing force or rite (such as the potlatch or pig cycle made famous in anthropology) that prevents social collapse or a mechanical explosion. The financial crisis that burst out in 2008 is grist on the mill of such theories that predict the fall of the West due to its inorganic political philosophy. The 'monotonic' idea of progress is not only self-defeating, but, says Dugin, also racist because it discriminates against both our past and 'peripheral' societies by considering all of them 'backward'. This, the claim of universality, is the third criticism and the motive for an anti-globalist and relativist philosophy, the 'Fourth Political Theory', that is forecasted to succeed the earlier three political theories which all proved to be failures: liberalism, communism, and fascism.

> Spiritually, globalization is the creation of a grand parody, the kingdom of the Antichrist. And the United States is the center of its expansion. American values pretend to be 'universal' ones. In reality, it is a new form of ideological aggression against the multiplicity of cultures and traditions still existing in the rest of the world. I am resolutely against Western values which are essentially modernist and postmodernist, and which are promulgated by the United States by force of arms or by obtrusion (Afghanistan, Iraq, Libya, and perhaps soon, Syria and Iran).[54]

---

[53] Alexander Herzen is never mentioned in Dugin's *Fourth Political Theory*.
[54] Alexander Dugin, 2012 (original 2009), *The Fourth Political Theory*, p. 114.

The 'Death of God' imperative is employed by Dugin as quintessence of the modernist and postmodernist void, hence his characterization of the forces of globalization as Antichrist. We will not go into the incongruity of this label when applied to parts of the world (like the US) that obviously know abundant religious activism. My aim is not to track logical obscurities in someone's thought, but rather to understand the thought's logic against the background of a geopolitical change. My goal jives perfectly with Dugin's own outlook – he considers the 'spatiality' of our existence a better guide to the destiny of peoples than some hollow vision of the future based on technological or economic growth. Russia's destiny is in Eurasia, a place where globalization does not square with the mind, or rather – in Heidegger's words – the 'being' (Dasein) of people.

This sounds great but does not answer the question of how the Fourth Political Theory eventually should guide political decisions (apart from a respect for tradition and ethnic diversity), particularly if there is a material crisis or dissatisfaction with life among people. Dugin wisely remarks, 'In order for Russia to save herself and others, creating some sort of technological miracle or a deceptive strategy is insufficient'. But he must postpone the revelation of new, alternative political strategies until the arrival of a magical event, a Russian *Ereignis* (Heidegger's Event), that will clear the road forward as so often happens in the ultimate depth of a crisis: 'This will be that very "Event", unique and extraordinary, for which many Russian people have lived and waited, from the birth of our nation to the coming arrival of the End of Days'. Thus, the distress of a people in a changing world again results in *millennialism* of a kind.

The religious character of these statements is not the tone of an isolated voice but echoes the viewpoints of the Russian Orthodox Church and its patriarch Kirill. The Church is advocating the idea of 'Holy Rus', which in Kirill's speeches alternatively consists of certain territories (nations) in the old Russian Empire or places where some religious event (like the establishment of a church) has taken place.[55] It is not a map but rather a dynamic conception of a progressing Christianization that crosses borders and nations (respecting sovereignties!) with a spiritual center in Moscow and other important satellites in historic places of the Orthodox cult (such as Kyiv). Its missionary message also points West, where the belief in individual freedom and tech-

---

[55] M.D. Suslov, 2015, 'Holy Rus': the geopolitical imagination in the contemporary Russian Orthodox Church.

nical progress has led to a 'spiritual colonization' of the world. There is a remarkable resonance between the Russian state and the Orthodox Church in the definition of security and spiritual security. The 'Russian Federation National Security Strategy' mentions next to dangers like natural disasters and terrorism the destruction of traditional Russian religious and moral values as a serious internal threat. [56]

'Black hole': Somalia

Among the three cases discussed in this chapter, Somalia is the only one in which collapse entailed the protracted breakdown of political authority in a large part of the former state. Somalia collapsed in the most absolute sense of the word, a fact which became very clear by the time the international community sponsored ill-fated efforts to promote conditions for the return of some form of central government (1993-1995). Indeed, the eyes of the world were focused on Somalia when US attempts to deal with Somali warlords (under the aegis of the UN (UNOSOM)) ended in disaster. TV-images of killed American soldiers dragged along through the streets of Mogadishu incited the American public to immediately reject any further contribution to peace missions in Africa. Somalia became the most tragic symbol of international impotence; regardless of the power of the intervening agency, a black hole engulfed each attempt at state-building without leaving any tracks (apart from further aggravating the situation). As Ken Menkhaus remarked about Somalia: 'It is not the *existence* [nor might one add *the absence*] of a functioning and effective central government which produces conflict, but rather the *process* of state-building which appears consistently to exacerbate instability and armed conflict'.[57]

On the question of what exactly caused the state to collapse in Somalia, a reasonable preliminary answer would be to say that it hardly got the time to develop as a state. It is doubtful if Somalia ever truly qualified as a full-blown state. The answer of course elicits a comparison with the development of other 'new' states: Norway, Finland, India, Indonesia, etc. These states all continued their existence without collapsing (even temporarily) and irrespec-

---

[56] Lucian N. Leustean, 2018, Eastern Orthodoxy, geopolitics and the 2016 'Holy and Great Synod of the Orthodox Church'.
[57] Ken Menkhaus, 2004, *Somalia: State Collapse and the Threat of Terrorrism*, p. 18.

tive of having been a colony. In a comparative analysis of the collapsing state in Lebanon, the former Yugoslavia and Somalia, Caty Clément enters into the issue of the acceptability of the label 'state' by providing a *minimal* definition: a state has a delimited territory, a coercive body capable of eliminating external and internal rivals and an administrative body that is at least capable of channelling resources towards the coercive body.[58] Although Somalia, according to this definition, qualified as a state after its independence in 1960, this definition also provides the key to answering the previous question about the causes of state collapse. The Somali pastoral mode of production provided extremely scant resources, which meant that the 'channelling efforts of the coercive body' were mainly directed at external support and consequently became dependent on capricious geopolitical conditions. No young European or Asian state had to function in similar conditions. However, one might argue that even the harshest environment might allow some frugal government to survive on internal resources. The inability to maintain even a weak government begs an explanation. This is also the assumption behind Clément extending her analysis to earlier critical and successful periods in the state-history of Lebanon, Yugoslavia, and Somalia. The fact that crises have previously put a political system to the test, without crossing the fatal breaking point, accentuates the occurrence of state collapse as something remarkable that needs explanation. Clément's QCA[59] analysis suggests that state collapse occurs if four crucial variables - international shift, economic shift, advanced group (people with economic and intellectual resources but no access to power) and elite renewal (no co-optation of representatives of diverse groups in leadership) - all score positive (i.e. a negative influence on social cohesion and political stability).[60] All these conditions occurred on the eve of the Somalian collapse.

The Somalian state emerged from colonial rule as an amalgamation of a British protectorate and an Italian colony. A substantial number of Somali's were left outside the new state territory: inhabitants of the French colony of Djibouti, pastoral tribes living in the Ogaden region of Ethiopia and agriculturalists in Kenya. The new government, fully aware of the slow and troublesome road to legitimacy that runs through the pursuance of economic

---

[58] Caty Clément, 2004.
[59] Quantitative Comparative Analysis (or Boolean analysis) searches for combinations of conditions that either assume the value of yes (present) or no (absent).
[60] Caty Clement, 2004, p. 18.

goals, saw a quicker way to mobilize popular support. It proclaimed the struggle for regaining the lost Somali territories as the national challenge. In 1963 war broke out with Ethiopia that, with no surprise in view of the great difference in military strength, Somalia lost. It was the heyday of the Cold War when superpowers were eager to score off their adversary by winning friends in countries of the Third World. Particularly the Horn of Africa attracted much attention because of its geographical location near the sea routes to the Middle East and as a suitable environment for military bases (such as the American Rapid Deployment Force). Somalia needed a huge military build up to catch up with Ethiopia's military strength. Only the Soviet Union was willing to meet such large demands. The national budget allocation to defence and security rose to 38 percent in 1969, implying an quick-growing military interest group.[61] Since those propagating the aim of 'greater Somalia' could count on wide support, this dangerously unbalanced development did not arouse serious political problems. Unfortunately, the then-president Egal, who steered in the direction of reconciliation with neighbouring countries, could not raise the same level of support. Notwithstanding the favourable domestic outcome of his peace politics, a military coup ended the civilian regime in 1969. This cleared the way for further military build-up under the regime of Siad Barre ('the Dictator') and an inevitable second war over Ogaden with Ethiopia in 1977-1978. Somalia was again defeated.

The disastrous outcome of the aggressive pursuit of a 'greater Somalia' incited the Soviet Union to cut off its military and economic aid to Somalia and shift its loyalty to more powerful Ethiopia. Despite this, Siad Barre and his military regime survived, mostly due to the willingness of other international actors (like the US) to fill in the void. The disillusioned Somali people were left with a bankrupt state and, perhaps even worse, an administration in which civil servants had never been accustomed to pursuing the public interest. In Somalia, the state bureaucracy was considered a venue to enrich oneself or one's kinsmen, or as a pool of rewards for loyal servants of the regime. Civil servants' revenues in poor countries may be so low that they are compelled to seek other paid jobs to supplement the household income. This does not bolster the efficacy of a bureaucracy. Anyhow, in Somalia most of the foreign assistance found its way to the state class, which contributed to the creation of a formal economy separated from an informal (shadow-)economy

---

[61] Abdullah A. Mohammad, 2002, *State Collapse and Post-Conflict Development in Africa. The Case of Somalia 1960-2001*.

built on market production by businesspeople or on remittances sent from abroad by Somali migrant workers. More than a million refugees, fleeing from Ogaden at the outbreak of war, caused further stress in the country's politics and economy.

Since Siad Barre's regime lost its legitimacy when it lost the war, it had to resort to methods of brutal repression to survive. Apart from arresting people and pursuing a scorched earth approach in the hometowns of dissidents, it proceeded by creating or reinforcing divisions between lineages or sub-clans.[62] By giving preferential treatment to certain sub-clans – such as providing them with money and weapons – the president hastened the transformation of the nation to an assembly of people that had lost all social cohesion. The logical result was that resistance movements made attempts to overthrow the regime. Upon failure to do so, the members of such movements fled to neighbouring countries where they continued their agitation. When the Barre regime, in an act of despair, signed a non-aggression pact with Ethiopia in 1988, thereby depriving the resistance movements of their military bases, the rebels started an all-or-nothing offensive in their homeland areas. Although the government responded with a full-out war effort, reducing a large city (Hargeisa) in the Isaac homeland to ruins, the result was that effective central authority ceased to exist. A scorched earth strategy in a domestic war means that there is no victor; it is queering one's own pitch. The Barre regime had only started a chain reaction leading to its own doom. As Abdullah Mohamoud writes: '…since that uprising [in 1988] the state began shrinking from its provinces while clan-based insurgencies were emerging one after another to take over the control of the abandoned peripheries. This means that the Somali state first collapsed in the peripheries before it finally exploded at the centre'.[63] The latter happened in 1991.

One of the reasons for the weakness of the military climaxing in the early 1990s – and not earlier when the regime had survived disasters like defeat in war – was the shift in the international political climate. The end of the Cold War meant the disappearance of superpowers bidding against each other to gain access to countries of the Third World, or to 'contain' the other superpower. Earlier foreign financial support, both military and economic, ceased to flow ('international shift') and could not be replaced by domestically-generated resources since the primary sector had suffered from years of

---

[62] Abdullah Mohammad, 2002, pp. 126-128.
[63] Ibid., p. 130.

neglect. Siad Barre was only able to attract foreign financial support by appealing to foreign donors on behalf of the refugees (probably inflating the size of the problem). The refugees, many of them Ogadenis belonging to Siad Barre's mother clan, became pawns in the divide and rule play of the president. They were allotted the better land and even armed to fight the (Isaac) clan that had to make room for them. So, the final attempts of the ruling regime to remain in power ultimately eliminated the last remnants of national cohesion and strengthened the unity and determination of adversaries.[64]

It is not difficult to point to causes explaining the fall of a regime, but that is different from explaining state collapse. After the ousting of the Dictator, the ruling Somalian elite disintegrated into clan-based warring factions headed by warlords. The most infamous name that emerged in the Western media was general Mohammed Farah Aideed, who the US media placed in the ranks of Saddam Hussein, Muammar Khadaffi and other leaders of 'rogue states'. However, the designation 'state' was wide off the mark in Somalia in the 1990s. After the disappearance of Siad Barre, Mogadishu 'civil society' appointed a businessperson, Ali Mahdi Mohammed, as interim president. This move was countered by one of the military leaders, General Aideed, who had a large role in ousting the Barre regime. Simply put, he considered himself the most rightful candidate for the presidency. The fighting that broke out divided the heart of the country and destroyed the capital Mogadishu. In the north the members of the Isaac clan, who in 1981 had founded one of the opposition parties with a base in Ethiopia, the Somali National Movement (SNM), declared themselves independent under the name Republic of Somaliland, another *de facto state*. In the south, civil war went on uninterruptedly following lines of solidarity defined by kinship and clan. In 2014 a large part of central and southern Somalia was in the hands of the Al-Shabaab Islamic militia. The Ethiopian military, the African Union (AMISOM) and the Somali Federal Government have all cleared Mogadishu and its surrounding areas.

It is tempting to point to the pervasive clan structure as one of the basic conditions that undermines the viability of the state in Somalia. Indeed, it is presented as a social ill in Nuruddin Farah's novel *Links* (2003). The book recounts the experiences of a Somali exile Jeebleh, a writer like Farah, visiting his homeland after spending years in the US. Upon strolling in the streets of ruined Mogadishu, he comes upon a scene of an epileptic man suffering from a seizure. A circle of people is staring at the man without lifting a finger for

---

[64] Ahmed I. Samatar, 1994, *The Somali Challenge. From Catastrophe to Renewal?*

helping him. On being asked why everybody is so passive, one of the onlookers says, 'We do not bother with people we do not know'. Whereupon Jeebleh shouts: 'Why do you need to know his clan family before you help him? What's wrong with you?' The existence of this type of 'wrong' in a society is certainly unsettling, but not necessarily prohibitive, in state building. Elsewhere, states have been able to partly get around cultural or ethnic splits or soften them without being put out of order.

The problem of any state-building effort in Somalia, as Menkhaus explains, is that so many local participants have veto-power over proposals for state-building or law enforcement.[65] Unsurprisingly, they do not hesitate to use it because they always fear that an opponent may get a better deal. Any Somalian considers central power as something that should be grabbed by his own group or (sub)clan. But this challenge is not the basic cause of war and disorder in Somalia. The problems that the international community experiences are largely the result of what they attempt to do: 'rebuilding' a set of institutions that never existed in Somalia. How then was law and order preserved in the period before 1991? The designation 'social contract', *heer* (or *xeer*) in Somali language, is more appropriate than the term law or police. It was an oral tradition of agreements about access to resources, water and pasture, and about the punishment of homicide and theft. Moreover, Islam provided rules (*sharia* law) that helped to maintain order in society. Islamic institutions owed their legitimacy to a successful organization of social care. Such institutions perpetuated a degree of order even where centralized control was largely absent. In present-day Africa, 'the central state is only one of a multitude of actors playing …roles' which people who live in the strong welfare states specifically attribute to the state.[66]

What then was the crucial step to a complete breakdown? According to Clément's (QCA) analysis, the missing link that completed the magic pattern of collapse (four negative conditions) was 'advanced group'. The increasing alienation of the Isaac clan from the ruling political (Hawiye) elite elicited the secession of an important economic region (Somaliland) and unleashed a state of anarchy from which Somalia could not recover. 'Identity', first as a dimension that prevented territorial collapse, finally caused the agony of a state. However, we should also realize that in Somalia there was no strong

---

[65] Ken Menkhaus, 2004, pp. 32-33.
[66] Ton Dietz and Dick Foeken, 2001, 'The crumbling of the African state system'. See also Chapter 7.

national identity that was superseded by clan identities. Identity did not keep the 'nation' together; the state (or the political elite) rather intensified clan divisions with its divisive partiality. In this way, (adverse) governance and identity were interacting to bring Somalia down. Foreign intervention only continued this mechanism, as it did in many other places of the world where the struggle of local tribes against the moral and economic pressure of central governments is mistaken for an epidemic of Al Qaeda terrorism.[67]

Re-territorialization within Somalia's boundaries
The poisoning of state legitimacy by clan prejudices could only clear the way for a re-territorialization along identity lines. The outcome was one territorial unit that aspires to be a state, Somaliland, and another, Puntland, that seeks autonomy in a future federal Somalia. Both derive their identity from the predominance of a specific clan (the Isaac in Somaliland and the Darood in Puntland) but also from their former status as British protectorate (Somaliland) and Italian colony (Puntland like the rest of Somalia). Next to these territories, de-territorialization has transformed into an unstable pattern of shifting alliances between families, sub-clans, and warlords. They are a black hole in the terms of Deleuze and Guattari (see quote at head of chapter) because they suck up all activity and accumulation without ever becoming a fixed point of reference in an external world.

Neither of the two more stable territorial units that emerged after 1991 have succeeded in becoming a fixed point of reference in the world of states: they are at best *de facto state*s. The difference between regular and *de facto* states is the simple fact of international recognition, which entails a host of difficulties if it is not granted. Authorities in Somaliland, who pride themselves on running an organized state with levels of safety that are not only higher than in Somalia but higher even than in neighbouring states like Kenya, repeatedly deplore the attitude of UN officials who only want to talk about development programs bracketed together with (southern) Somalia or the so-called 'Transitional National Government (2000-2004) of Somalia'. In trying to reach an agreement on UN funding of school textbooks, the UN secretary general's representative Winston Tubman – a Liberian statesman – came with a disconcerting note for Somalilanders. As an incensed comment in the Somaliland Times ran:

---

[67] Akbar Ahmed, 2013, *The Thistle and the Drone. How America's War on Terror became a Global War on Tribal Islam*.

Somalilanders who are used to unreasonable demands and claims by overpaid and underachieving UN bureaucrats, were shocked by the extent of mean-spiritedness and hostility shown by Mr. Tubman towards them, as evidenced by his letter of 21 October 2003 which called for the removal from the Somaliland curriculum of:
- all topics related to the historical background of how Somaliland regained its independence on May 1991;
- all the maps showing international boundaries between Somaliland and its neighbouring countries as well as such purely physical features as hills and mountains.[68]

According to the Somaliland Times editorial, 'Mr. Tubman ostensibly wants to protect [school kids] from "secessionist infection"'. The fact that Somaliland actively promotes school attendance, organizes free parliamentary elections and even collects housing data for town planning in Hargeisa can hardly be ignored by UN officials who are accustomed to never-ending stories about state failure in Africa. A UN Joint Needs Assessment team and the World Bank struggled to assure local officials that assistance to Somaliland would occur separately from Somalia, while at the same time working in an international framework that labels Somaliland 'Northwest Somalia Region' and designates the Transitional Federal Government (2004-2012) as the single Somali authority that signs agreements with international actors.[69]

Why is recognition forbidden action even when seceded units are politically more coherent than recognized but failing states? One argument is the foundation of the United Nations Organization as a protector of the sovereignty of its members. While accepting external intervention for the sake of human rights, secession, often motivated by appealing to the same UN Human Rights Charter, is not on the UN agenda.

Another argument against recognition is mistrust in the domestic legitimacy or future stability of a breakaway state. If ethnic tension is an important motivation in secession, newly created territorial units will very often have to deal with minorities again: the problems in the newly independent Soviet Republics are illustrative. Even Somaliland is inhabited by clans other

---

[68] 'It's our curriculum', *Somaliland Times*, 15 February 2004 (http://somalilandtimes.net/-2003/107/10700.shtml). The reference to hills and mountains is curious but see the political message of maps created in the aftermath of the Pyrenees Peace Treaty (Figure 3.3).
[69] 'The World Bank and the UN in Somaliland', *Somaliland Times*, 21 January 2006 (http://somalilandtimes.net/sl/2005/209/01.shtml).

than the Isaac. If new polities can overcome ethnic differences, why not ask to make the same effort in the larger territorial framework (Somalia)? Whereas secessionists are appealing to the new domestic peace they have secured, the UN fears an infinite chain of territorial divisions. UN officials rather see a challenge in the reverse direction: linking different groups in one cooperative framework rather than erecting boundaries between people.

A third argument is 'geopolitical': Islamic countries like Egypt oppose the division of Somalia, which is considered a potential ally against Ethiopia, a 'foreign' Christian body in a predominantly Muslim region. Western countries would not like to offend the Arab world by insisting on the independence of Somaliland.[70]

Notwithstanding international adversity, Somaliland unleashed a new 'diplomatic offensive'. It proceeds along two courses: international and transnational. One of the members of a parliamentary delegation to neighbouring countries declared early in 2007: 'Previously our contacts were limited to government-to-government, but we have to raise them to the level of parliaments, parties and people' (i.e. transnational).[71] Since that time more official agreements (even involving recognition of Somaliland passports) have been achieved. The US, UK and the EU have implemented official ties and there are foreign and/or Somaliland offices at home and in countries like Kenya, Ethiopia, South-Africa, Djibouti, France and Sweden.[72] The presence of 'International Partners' of Somaliland may also act as a watchdog in safeguarding democratic procedures. For example, the Partners issued a warning about the postponement of presidential and parliamentarian elections in early 2017 (attributed to the consequences of severe drought) which emphasized that Somaliland's lower house will have been sitting for 12 years.[73] On the other hand the establishment in 2018 of foreign military bases, like the one of the United Arab Emirates in Berbera and possibly one of Russia, might involve Somaliland more in geopolitical conflicts than would be healthy for a non-recognized state.

---

[70] Gérard Prunier, 2010, 'Le Somaliland, une exception africaine'. *Le Monde diplomatique*, October 2010, p. 6.
[71] 'Somaliland launches new diplomatic offensive', *The Somaliland Times*, 27 January 2007.
[72] See http://recognition.somalilandgov.com
[73] Statement by international partners to Somaliland on the postponement of the Presidential and Local elections. https://eeas.europa.eu/sites/eeas/files/final._statement_by_the_international_partners_to_somaliland._2017.01.26.pdf

A Somali answer to collapse in a globalized setting?
It is undeniable that the collapse of the Somali state, while induced by globalization, also unleashed a dramatic process of globalization, whether through the influence of the Somali diaspora (which amounts to more than 15% of its indigenous population) or through the interventions by foreign powers and organizations within its boundaries. One of those organizations, Al Shabab, has a distinct religious character and, while it claims to represent the interests of the Somali people, it is also linked to such transnational movements as Al Qaeda. It reaped notoriety by bomb attacks against the Transitional Federal Government (TFG) in Mogadishu and against targets in neighbouring countries (Kenya, Uganda) where the governments support the internationally recognized TFG. In October 2017 an attack of Al Shabab in Mogadishu took the lives of almost 600 people.

Among the many Somali political movements that struggled for central power in the mid-2000s, there was one, the Union of Islamic Courts (UIC), that kept aloof of the struggle between warlords. With its non-particularistic political programme and emphasis on creating stability and security, the UIC won broad support and eventually succeeded in chasing the TFG from Mogadishu in 2006. However, the court's strict application of Sharia law in specific districts spread discord among supporters and evoked the Taleban nightmare among foreign powers like the United States. They did not hesitate to support an Ethiopian military intervention that within a few days succeeded in expelling the UIC from Mogadishu. Frustratingly, it is precisely such foreign intervention that inspires certain break-away groups from the UIC (like Al Shabab) to raise the internal struggle in Somalia to a transnational issue and a religious war (against Christians). As one of Al Shahab's leaders commented after their bombing attack in Nairobi (2013): "The Mujahideen fighters refused to accept the [Western] invasion of their Muslim lands". But is this example of 'global framing' the spiritual upshot of a shock caused by globalization in Somalia?

The rise of global framing in situations where security and local ties have collapsed may result from two mechanisms: 1. Cultural resonance between transnational ideas and local traditions, and 2. The strategic calculation of local 'entrepreneurs' who seize on transnational movements to strengthen their own position. Peter Henne concludes that there is a lack of resonance between Somali culture and the Salafist ideology of transnational Muslim movements.[74] Somali Islam is rather moderate and the strength of tribal identities 'limits the appeal of the global Islamic identity championed by groups like Al Qaeda'. However, there are local entrepreneurs 'willing to advance the transnational claims of groups like Al Qaeda'. Such local entrepreneurs do not offer a stable and consensual framework for thinking about Somalia's territorial unity and place in the world: this ideal evaporated from the Somali intellectual horizon with the post-colonial disillusionment about the national state. In the early years of independence, it was easy to see colonialism as the mother of all political ailments in Africa, as a foreign force that had repressed all natural and territorial African entities. African writers considered the liberation of the nation an essential precondition for the development of an 'authentic' African narrative.[75] After the experience of so much abuse of power by local authorities and failed liberation struggles against fellow Africans, the mood turned inward to self-torturing thoughts about one's identity.

> **Box 6.3. Nuruddin Farah's Somali maps**
>
> Nuruddin Farah's novel *Maps* (1986), a tale with a telling title, starts with the almost insane musings of its central figure, Askar, about the continuity between his body and the environment both in a social and physical way. Taken as an infant from the body of his dead (Somali) mother in the Ogaden region, he is brought up by a woman, Misra, with a vague Oromo/Ethiopian identity. A more intense conflict along the dimensions of corporeal and geographical identity is hardly imaginable. When the Ogaden war breaks out (1977), Askar, at the age of seven, is received in the household of his cosmopolitan uncle Hillaal and his wife Salaado in Mogadishu. The intellectual climate in which he grows up from now on provides ample space for torturing questions about his personal loyalty (to Misra, the alternate 'mother' who saved him as infant) and the call of national identity, materialized in political and ethnic maps. Is there an absolute truth in the will of an entire nation? If so, this truth would demand that Askar join the Western Somali Liberation Front, and he yearns to do that. Although uncle Hilaal once stated that it is impossible to prove an entire people (in contrast to an individual) 'to be terribly mistaken', as an intellectual he is rather worried about ethnocentrism and the fact that a

---

[74] Peter S. Henne, 2013, 'Is all jihad local? Transnational contention and political violence in Somalia'.
[75] Simon Gikandi, 2012, 'The politics and poetics of national formation. Recent African writing and Maps', p. 451.

> people can all the same be 'sadly mistaken about themselves'. Such musings are inspired by the fact that Somalia is not a casual agglomeration of different ethnic groups, like most other African states, but a state that has been cut off from its people. Yet the logic of the ethnic map cannot be restored without war or without accepting that the claimed territories are not ethnically pure. The author ends his novel with a scene in a police station where Askar wrestles to document his identity, clearly haunted by the territorial and ethnic conundrum.

Farah's novel was written before the rise of aggressive Muslim transnationalism and illustrates the obsession with closure and identity that characterized the territorial strategy of the Siad Barre era. But as the novel suggests, maps only reinforce the feelings of perplexity in the personal sphere. Since the international organizations' 'good governance' programs also result in corruption and tribal competition (notwithstanding positive examples like Somaliland), we can only conclude that there is reasonable prospect of reterritorialization. However, there is a gleam of hope in the place that was last touched by the forces of disintegration, but that always kept a spirit of cosmopolitanism: Mogadishu. In the past decade a stream of people from the diaspora, among them many who enjoyed relative economic success in Europe or America, have returned to Mogadishu to try their luck in 'a city on the upswing'.[76] They defy the still-real danger of terror attacks by believing that their experience living in racially mixed societies where clans count for nothing will contribute to the regeneration of the country. However, it is precisely their being different which constitutes a new problem. They invoke envy by occupying scarce and desirable jobs while not having suffered the hardships of those who remained in Somalia. In a way they threaten to become just a new clan adding to the existing tribal diversity.[77] Yet, if they succeed in establishing institutions, for example in the hitherto untouchable spheres of banking and justice, they will introduce a moral principle into the process of globalization that has been thus far only known for its destructive power.

---

[76] Andrew Harding, 2016, *The Mayor of Mogadishu*, p. 219.
[77] Ibid., p. 232.

Conclusion: the re-territorializing drive after state collapse

Did 'the music of our time' unleash the powers of de-territorialization in the cases of state death discussed in this chapter? Did they show how an innocent deed like singing could suddenly become a tool of destruction? The power of music, with its mystic influence, woke Estonians up to the link between their nation and the rest of the world. Though powerful, this was not sufficient as such to cause the break-up of the Soviet Empire. A chain of other events and structures was necessary: the unanticipated reaction of government agencies that put freedom of expression on the agenda, the political and sometimes violent demonstrations in other places (with victims of military action) and of course the federal structure itself of the Soviet Union, which promised freedom of secession (although it was never meant to be implemented). The line of flight went over identity, but the nature of Soviet control determined its final landing (re-territorialization).

In post-Soviet Russia, the music of a new 'information age' promised access to the world, whatever the power or size of one's polity. This unfortunately turned sour when the expected wealth for ordinary citizens failed to materialize all while Russia's influence in the world appeared to diminish. By continuing its 'anti-imperialist' stance and adopting an anti-globalist emphasis on ethnic diversity (akin to many extreme right movements in the West), Russia offered a solution that claimed a leadership role in the world while at the same time satisfying the domestic need for a moral opposition of hegemonic liberalism. After all, the information age has in actuality brought less of an industrial upswing in Russia than a better understanding of the importance of public opinion. This is manifest by Russia's incessant attempts to foster populist discontent in the West by manipulating social media platforms by means of cyber-war. In this sense, the new Russia has adopted 'governance' features associated (in the preceding chapter) with globalization or late-modern territoriality.

This also pertains to the penetration of the state by non-official or illegal actors as well as the porosity of borders (closure). However, Russia's reliance on the transnational institution of the Orthodox Church combined with the state's proclamation that protecting Russian values is a matter of state security suggest an active role of identity which does not conform to our ideal-type of globalization. Rather, Russia's actions and policies evoke the imperial model, which also defines borders as zones of instability and attempts to unify people in a struggle against external barbarians. This model also advocates respect for internal diversity, which blurs the real lines of power and influence. The new states of Russia's 'near abroad', those that share the Orthodox heritage, struggle to a different degree with this transnational influence. Ukraine's president Poroshenko aspired to create a Ukrainian national Orthodox church, and the link between Georgia's Orthodox churches and the Moscow Patriarchate irritates Georgia's political leaders. Conversely, other former Soviet nations have wilfully accepted the transnational influence of Russia because it suits their economic structure and the illiberal urge of their leaders.

The defeat of France in the Second World War was sufficiently shocking to evoke significantly more than a desire to restore the old pre-war state. There was a widespread conviction that France had gone astray, not merely militarily but in a moral sense, as the Vichy regime emphasized. And the music of the time did not suggest that a return to normalcy would ever be possible; the world had changed too much, almost as in a geological process. In the words of Claude Jamet: 'The frontiers jump, the map moves, the world transforms before our eyes' – and not merely because of Hitler's designs, but as an expression of some 'innate' geopolitical force. The 'new France' demanded a transnational vision, and even when the nightmare of German occupation ended this idea of a differently interconnected world persisted. It was preserved in the people's expectations for a European community, in the cosmopolitanism of the existentialists and – despite his state-centred outlook – in the courage of President De Gaulle to put an end to the colonial relations with Algeria (1962). The post-war influence of American values (and political pressure) cannot be denied, but it all the same belonged to the unique music of the time, which sent the message that most European countries were too small to cope with 20th century war and its aftermath. This was obviously expressed in the integrative moves of France's small neighbours, the Benelux countries. The significance of occupation and defeat in war is also demonstrated (in the opposite sense) by the ultimate failure of the United Kingdom to continue its membership of the European Union. The rise of supra-national institutions in the spheres of banking, justice, and financial support, however, is anyhow a clean break with the state-centred tradition.

The civil war that wiped out any and all effective government in Somalia was a direct consequence of Siad Barre's policy of advancing his own group (Maherteen) and pitting clans against each other. The tradition of reaching broad support by informal arrangements between (sub)clans was abandoned precisely because the Barre's regime could rely on external financial support, which was easily available for states that held a strategic position in the Cold War era. The line of flight obviously ran over the family and the clan, but since the national road to making 'the self strong and whole' (words of uncle Hilaal in Farah's *Maps*) had proven a failure, nothing else seemed to remain but two options: either retreating to the tribe or embracing a global identity such as religion. However, there was a literal line of flight in the shape of a diaspora, which was of import due to remittances sent in from abroad that bolstered the economy and the introduction of a spirit of enterprise when members of the diaspora returned to Mogadishu. Its members also showed a commitment to good governance in Somalia, and particularly in local government (they have a distaste for national-level politics).[78] Here, the trend towards 'glocalization' is obvious, but so is the awareness that enterprise is a matter of community building. It is too early to judge its success, but its experience with cross-tribal and cross-border solidarity may give the diaspora a clear lead.

Of course, Somali*land* is an example of reterritorialization which (as yet) seems to offer stability without resorting to a mobilizing vision like religion or an enemy. Explanations are not difficult to find: its history as a British protectorate with a tradition of self-government, its relative homogeneity (dominant Isaac clan, pastoralist way of life), etc.[79] A less obvious reason is precisely the *absence* of any official international support which brought home to Somalilanders that they were themselves responsible for a stable political settlement. No internal struggle about foreign funds or (its opposite) external requirements on the distribution of funds. As Sarah Phillips has argued President Egal's choice of entrusting trade to a small business elite both helped the economy and jumped over more traditional oppositions such as between clan elders.[80] It also fitted the trend to give non-traditional and non-institutional voices a more prominent place. In admitting a wider range of voices

---

[78] Laura Hammond et al., 2012, *Cash and Compassion: The Role of the Somali Diaspora in Relief, Development, and Peace-building*, p. 91.
[79] Michael Walls and Steve Kibble, 2013, 'Identity, stability and the state in Somaliland: indigenous forms and external interventions'. See also Hammond et al, 2012.
[80] Sarah G. Phillips, 2016, "When less was more: external assistance and the political settlement in Somaliland'.

than those of clan representatives or established powers (warlords), this form of governance also appeals to a different mind-set. As we have seen, the officers for Somaliland foreign relations have already been compelled to launch a 'transnational' diplomatic offensive ('from people to people'), but this also had a counterpart on the domestic scene. One group that is traditionally excluded from the political process in Africa is women. During one of the conferences about the future of Somaliland in Borama (1993), women, although they could not vote, organized themselves as a pressure group that barred the delegates from leaving the conference venue until they made some important decisions.[81] Since such groups often have bonds (support and information) with people of the same kind elsewhere in the world, globalization (in the shape of transnationality) here enters the picture as a factor in re-territorialization. Whether one would speak of a spiritual reorientation is a matter of taste. Nevertheless, it is certain that this re-territorialization involves a moral readjustment among groups that, while fearing the loss of their influence, hope for the possible reward of stability.

Table 6.1 The role of territorial dimensions in reterritorialization after state collapse

	France (1945)	Soviet Union (1991)	Somalia (1991)
Regime basics	Governance Identity	Governance Closure	Closure Identity
Line of flight	Closure	Identity	Identiy Governance
Reterritorializing vision	Vichy: Govern./Identity EU: Govern. /Closure	Russia: Identity Closure	Somaliland: Governance Closure

The three different conditions of state collapse are represented in the territorial language used in this book. Table 6.1 shows the initial position of France as a typical product of the European high-modern age (governance-identity), the Soviet-Union rather as something imperial (governance-closure) and Somalia as an early-modern state struggling to reconcile its social nature with the map (governance-identity). While it is difficult to associate the de-territo-

---

[81] Michael Walls and Steve Kibble, *Identity, stability etc.* 2013. *After Borama: Consensus, Representation and Parliament in Somaliland.* Africa Research Institute.
(http://www.africaresearchinstitute.org/newsite/wp-content/uploads/2013/05/PV-After-Borama-HR-for-website.pdf)

rializing urge (line of flight) with a single territorial factor, the turbulence of closure in occupied France and of identity in the Soviet Union is obvious. In Somalia, the combination of governance (distribution of resources) with identity (clans) turned out to be disastrous. More interesting, however, is the starting question: does the process of re-territorializaton after collapse show an adaptation to a new era? The evolution of France into a member of a supranational body is a clear indication. The continuation of an empire, the Soviet Union, in the shape of the Russian Federation, focused on transnational (spiritual) influence in and across its periphery, evokes the model of a foregone era. But in order to succeed, current Russian political action needs an involvement with international thought and values that classic empires would reject as superfluous. Military power, so essential in classic empires, is still a factor, but international relations are currently also controlled by another force: symbolic multipliers (see p. 127). Finally, Somalia's agony has not yet ended, but there are some developments that may prove self-reinforcing simply because there is no alternative. One is the role of the diaspora returning to Mogadishu. Their businesses may arouse envy, but they also attest the beneficial results of glocalization and set an example for others to follow. Glocalization means jumping from the local level, across the state level with its heritage of dictatorship and territorial antagonisms and ending at the global level. This is the reality in Somaliland, where from sheer necessity international relations occur in the shape of transnational contact from group to group. What seems a handicap now may prove to be an asset in the changing territorial order of the world.

Dying states demonstrate features of a globalizing world, but the picture provided above is far from complete, both in terms of results and suitable cases. Another complication is that a (territorial) 'hardware' change elicits moral reactions that, although may encourage people in the private sphere, may also create a backlash when political leaders exploit such ideas (perhaps for their own benefit) to change the whole territorial principle.

Chapter 7
GLOBALIZATION AND ITS DETRACTORS

> We've been traveling over rocky ground….
> You use your muscle and your mind and you pray your best
> That your best is good enough, the Lord will do the rest
> You raise your children and you teach them to walk straight and sure
> You pray that hard times, hard times come no more
> You try to sleep you toss and turn the bottom's dropping out
> Where you once had faith now there's only doubt
> You pray for guidance only silence now meets your prayers
> The morning breaks, you awake, but no one's there
> There's a new day coming….
> **Bruce Springsteen,** *Rocky Ground* (2012)

The bottom is dropping out

In a song that owes its musical style to the genres of gospel and reggae, Bruce Springsteen hints at some powerful images evoked in the Bible, such as Jesus expelling the money changers from the temple[1], the good shepherd and the seed that will not germinate when it falls on 'rocky ground'. It is difficult to ignore a reference in this song to the plight of the American lower and middle classes, who have not seen much fruit for their toil during the four decades that are often considered the beginning of the contemporary epoch of globalization.[2] No fruit at least in the sense of an American dream, in which hard work is rewarded and each generation is thought to do better than the previous one. The culprit has been earmarked as 'globalization' by anti-globalist demonstrators at the WTO conference in Seattle (1999) and the Occupy Wall-Street movement (2011). The anti-globalist sentiment culminated in the Republican conversion to unilateralism instigated by Donald Trump in 2016. But the definition of the global evil has shifted during this period from inequality and ecological threats on a global scale (the Seattle demonstrators were rather 'alternative' globalists) to the lack of democratic control in the

---

[1] Not in the fragment reproduced at head of this chapter.
[2] In his autobiography, Springsteen notes: *"I'd been following and writing about America's post-industrial trauma, the killing of our manufacturing presence and working class, for thirty-five years"*. Bruce Springsteen, 2016, *Born to Run*, pp. 468-469.

multinational system of big business and finance (Occupy Wall-Street and its more than 1500 parallel manifestations worldwide) and finally to Trump's political earthquake, appealing to nationalist and xenophobic sentiments. Together they reveal three basic shocks that globalization has caused: persistent inequality, democratic deficit and ethnic competition. It was not entirely economic; the event of 9/11 2001 contributed its own shock and imagery. Neither are the sentiments exclusively American because such threats are also felt elsewhere, for example in Europe.

The protests in Seattle targeted the spectre of a 'race to the bottom' in which the elimination of trade barriers would erode all progress in food safety, environmental protection and healthy working conditions that had been achieved after long political struggles in the US and Europe. How easy would it be for shady international companies (or Western industries that move production stages abroad) to force out such socially accountable activities when they don't have to account locally for social and environmental costs. This was not merely a self-serving argument, as the protesters also argued against degradation of the environment in the Global South (which, of course, touches all of us in the shape of global warming). The belated shock – after more than a decade – among Western fashion industries over the collapsing of factory buildings in Savar, Bangladesh (which killed about 1200 people in 2013) has also put the Seattle protesters in the right concerning working conditions in the notorious global 'sweatshops'.

The Occupy Wall Street movement did not voice clear political demands; it was rather an occupation of public space with the aim to shape a moral and autonomous community by representing the '99% against the 1%' with (financial) power. The movement was a direct response to the 2008 collapse of financial institutions and mortgage banks, which resulted in the wave of foreclosures on the housing market. With their primitive settlement on the tarmac of Zuccotti Park (New York) and elsewhere, they demonstrated the importance of face-to-face contact and the sharing of essential life conditions (like shelter, food, and maintenance) when forming opinions. There was a clear anarchist drive behind these protests – a longing to solve the basic problems of life without appealing to some strong power or organizational shift. This anarchist drive, at the same time, is one of the reasons for the decline of such movements and perhaps for the success of subsequent political entrepreneurs who have tapped the same frustrations against an 'elite' but who, like Donald Trump, hold onto the prospect of a strong, 'safe' national state and leader. A further reflection on the mechanisms behind the current production of wealth quickly reveals that such geo-economic power play is just as ineffective at solving the domestic problem of economic inequality.

Since 1990, the G7's relative share in the world's economic production has declined, and this has particularly hit the industrial working force. Whereas the industrial revolution, by reducing transport costs, turned these countries into global exporters in the 19th century (with the 'Great Divergence' between the industrialized world and the rest as result), the current transformation should be rather called the "Great Convergence".[3] Globalization has produced a rapid spread of knowledge; there is more industry that depends on skill and high-tech input, and the cost of communication over long distances has strongly diminished. The first consequence was the transfer of entire labour-intensive industries to areas with cheap labour and in its wake the outsourcing of production *stages*, including technical and managerial knowledge to low-wage countries. It means that the entire production cycle of a firm is scattered across the globe. Such is the case of computer firm Apple, which concentrates its research and service activities in the US, whereas most of the smartphones or parts are made in China. Protectionist measures like high import tariffs will not force the relocation of such production to the US because the firm would then lose its competitive edge in the global market. Moreover, protection would raise prices for consumers in the US and damage a substantial part of the national economy that is dependent on import. This is in addition to the costs of a trade war started by the Trump administration in 2018. Because there are other causes of weakening industrial employment (for example automation), it is likely that such strategies will lead to nowhere, which increases the call for a (moral) reorientation of the kind that has marked the great territorial shifts in history.

There is dissatisfaction with globalization in other parts of the world, which usually amounts to the feeling that inter- or transnational agreements have a negative impact on sacred or profitable local practices. Regulation on the intake of migrants unleashed popular resistance against the supranational institutions of the European Union, whereas the issue of international quality certificates has tainted globalization in the Global South as an instrument of unfair competition or worse, neo-colonialism.

---

[3] Richard Baldwin, 2016, *The great Convergence. Information, Technology and the New Globalization*.

The European Union

If there is a part of the world where globalization has been allowed to indent deeply on national sovereignty, it is Europe. However, the EU's increasing influence over rules and administration and a steadily widening membership did not arouse alarm until the middle of the new millennium's first decade. It started with referendums in the Netherlands and France that rejected the proposal of a European constitution (2005). Then came the financial crisis, which badly shook the economies of the southern member states, particularly Cyprus, Greece, Spain and Portugal. Since they were members of the common Euro currency area and not able to adjust their exchange rates in order to benefit from lower price levels, they were dependent on the European Central Bank for supporting their government budgets. The relatively low rate of internal labour migration in the EU (because of language differences) also did not help to alleviate unemployment problems in the South. Even worse was the growing antagonism between certain countries of the North, who accused the South of irresponsible budget policies, and countries of the South, who complained about unreasonably high interest rates that would only benefit banks of the wealthy member states (particularly Germany). The economic upturn of the second decade was hardly given the chance to ease these tensions because another challenge descended upon Europe: a refugee crisis caused by civil wars in the Middle East and failing states and economies in Africa. They came by the hundreds of thousands across the Mediterranean and Turkey, aided by a whole army of frontier-runner entrepreneurs. Tensions about admitting refugees emerged between Eastern and Western members of the EU. Scepticism about EU membership increased in all countries. EU membership seemed to make them susceptible to all kinds of disasters rather than strengthening their economic position in the world.

The situation of EU members formerly belonging to the communist bloc is even more precarious, because they emerged from an international constellation that already stressed transnational (communist) unity at the expense of national identity. They entered the EU with the assumption that they would be free to develop their national fortes within a protective shell that would fit in with their national (and Christian) tradition. To be sure, the EU political elite have always emphatically denied that there is an opposition between national identity and a unified Europe. On the contrary, protection of national (territorial) cultures is one of the basic aims in the European Charter of Fundamental Rights: 'The Union contributes to the preservation and the development of [its] common values while respecting the diversity of the cultures and traditions of the peoples of Europe *as well as the national identities of Member States*' (Preamble). Yet, a strong feeling of national pride is sometimes difficult to reconcile with the loss of sovereignty, and the latter is difficult to accept when economic advantages are unclear or Union-wide solidarity requires 'unreasonable' sacrifices. The manifestation of a severe brain-drain in the new member states is also a slur on national pride. There is a stark contrast between the global view inherited from the colonial experience of several West-European countries and the historic experience of the age-long struggle against 'barbaric' invasions from the East. This is a clash between deep-rooted geopolitical visions in which globalization and liberalism constitute one side and fear of foreign invasion the other side. The territorial shock involved in the latter seems to result in nationalism and a strong illiberal form of governance (Hungary, Poland).

There is fear as well among the longer-established EU members of contamination by illiberal practices (corruption, lawlessness) in new or prospective EU members. A Dutch referendum about the EU Association Agreement with Ukraine (2016) stipulated that the treaty would not automatically grant EU membership to Ukraine. 'Populist' parties in several EU states exploited these feelings with a movement that held up the prospect of leaving the EU after the example of Brexit. Yet, the sentiment in the UK cannot be compared with that on the continent, where the awareness of mutual dependence is stronger. There are other resentments resulting from the consequences of EU integration that are comparable to the disconcerting experience of US industry relocating to Mexico (or other effects of NAFTA arrangements in North America). Many Western European industries, the car industry as an outstanding example, have shifted production stages to the Eastern members like Poland, Slovakia and Romania, where wages are still lower. Countries where many jobs had been created in service sectors, like road transport, also saw employment drain away to the Eastern European member states. Trade agreements with countries outside the EU (which are all negotiated at the EU level) like TTIP and CETA[4] hamper the establishment of consumer or environmental protection measures because investors, who feel disadvantaged by national legislation, may lodge heavy claims for loss of profit. The settling of such disputes occurs through private arbitration

---

[4] TTIP – Transatlantic Trade and Investment Partnership (Between Europe and US), CETA – Comprehensive Economic and Trade Agreement (between Canada and Europe).

outside the reach of democratic control and without the possibility of appeal. Here we arrive back in Seattle, at the resistance movement against the WTO arrangements, where democratic control and protection of the environment constituted basic arguments.

The Global South

As the concept of a 'Great Convergence' reveals, globalization has improved the economic prospects of many countries in the world outside the West and Japan. Globalization has diminished *global* inequality, but it is also occasionally met with resistance, as demonstrated by two conflicts in countries of the 'Global South'. One is the post-colonial sensitivity for infringement on a state's sovereignty. Since globalization entails the advancement of general standards and rights, these may be experienced as colonial tyranny in a new guise. Second, globalization may erode traditional ways of life, even where they might be able to prosper in a globalized setting. In such cases globalization possibly offers strategies of resistance as well.

An example of (post-)colonial sensitivities aroused by transnational influences is the eco-certification of agricultural or fishery activities.[5] Consumer groups and ecological activists in North America and Europe have exerted pressure to guarantee food safety, but they also aim to ensure that production conforms to requirements of sustainability and decent labour conditions. Additionally, they work to eliminate natural and economic threats, for example by instituting measures to protect a coastline dependant on the preservation of mangrove forests. The FSC (Forest Stewardship Council) and MSC (Marine Stewardship Council) are well-known certificates that aim to protect the forests, along with their traditional ways of life and the sustainability of their food source (seafood). Where a primary activity has a relatively great weight in the national economy, like shrimp aquaculture in Thailand (the world's leading producer), such transnational concern may be perceived as infringement on national sovereignty. Transnational certification movements tend to see local government laws and regulations as inadequate, poorly enforced and corrupted. Conversely, Thai government representatives see such movements as an imperial strategy to create a level playing field that

---

[5] Peter Vandergeest and Anusorn Unno, 2012, 'A new Extraterritoriality? Aquaculture certification, sovereignty and Empire'.

primarily advantages foreign producers. Ultimately, a consensus has grown that attempts to find middle ground on the responsibility of local governments for certification contrasted with the voice of the consumers and their ecological concern.

Globalization may also turn up in the shape of multinational capital, which easily acquires a hegemonic role in countries where strengthening of the labour market is an urgent need. In such conditions, multinational projects may seem to have democratic consent from the outset. However, they are not always welcome in local communities. In the case of a mining project in Tambogrande (Peru) by an international firm, Manhattan Mining, the local population opposed plans to establish an opencast mine 3 km in diameter which would require the relocation of 8000 people.[6] Here national legitimacy, to which the firm appealed, was in contradiction with *local* legitimacy. As Haarstad and Fløysand explain, local opposition succeeded because the activists also managed to enlist (trans)national forces: NGOs who defended the right to protect a local agricultural way of life. In addition, the nature of the local production, the cultivation of lemons and mangoes, was successfully associated with Peruvian national identity and its traditional lemon-based dishes. Globalization may be a threat, but it also offers the possibility to 're-scale' narratives in defence of such threats. Such rescaling offers a recourse to what has been called 'glocalization' in a different context (see below). One may wonder if such examples offer a glimpse at the discursive and mental strategies required to adapt to the forces of globalization – without ever being able to undo the worldwide change that befalls us.

A new moral challenge?

Obviously, there are many evils in globalization – real or imagined – that elicit the political reflex of withdrawal to a safe national haven, often by erecting walls in the shape of tariffs or entry restrictions. The principle behind new territorial orders, however, is that such retrograde movements are not entirely feasible on penalty of economic or political damage. The negative effects of import tariffs have been mentioned, and the rejection of multilateral cooperation may ultimately become a security risk. We have also seen that protests

---

[6] Håvard Haarstad and Arnt Fløysand, 2007, 'Globalization and the power of rescaled narratives: A case of opposition to mining in Tambogrande, Peru '.

Figure 7.1. A documentary film about the sucessful struggle against the power a of a multinational enterprise in Tambogrande.

against a multinational enterprise in Peru were successful, but at the same time they mobilised support of international movements (NGOs). This reveals the complexity of a new globalized condition which can be condensed in the term 'glocalization'. The sequel of urban manifestations against global regimes like neo-liberalism or multinational capital (Seattle, Occupy etc.) is another example. Many authors have commented upon the lack of endurance and political response of these movements in the West – even more so since recent economic developments seemed to put the early 'alter-globalizationists' in the right.[7] To add insult to injury, as one author claimed: 'implosion of neo-liberalism in 2008 resulted in, paradoxically, more neo-liberalism'.[8] Several reasons have been suggested to explain this state of affairs: the purposeful lack of central leadership and political strategy in (neo)-anarchist action, the easy neutralization of street action by deploying police forces and the rise of new fears about security after the attacks of 9/11 (when terrorism

---

[7] Noah Smith, 2014, 'The dark side of globalization: why Seattle's 1999 protesters were right'.
[8] Blair Taylor, 2013, 'From alterglobalization to Occupy Wall Street. Neoanarchism and the new spirit of the left'.

became the overarching threat). There is another explanation which might account for the draining of energy from the struggle against neoliberalism: its relative success. Many of the ideals of the alter-globalist and anti-capitalist movements, such as the sourcing of ethical goods (respecting human and animal life and sustainability), renewable energy, informality and adapting to personal talent in the workplace, have become keys to commercial success and guiding principles in management thinking. Both the impossibility of creating a utopian community with alternative products (which the free market can produce for much cheaper) and the agility of the information society to make lifestyles marketable have diluted criticism on global capitalism. Neo-anarchism and neo-liberalism mingle easily. It is another example of the subtle ways in which a new age penetrates one's life and values without eliminating some of its problematic features, like multinational capital. Do we see here the outline of a morality that softens the impact of the globalized territorial order? Let us first look at the more conspicuous 'moral' attacks on globalization.

The historic chapters that constituted the first part of this book revealed two types of moral adaptation to fundamental territorial changes: **religion** and **nationalism**. Religion helped people to accept a new territorial power, an empire or a sovereign state, as an instrument of divine purpose. A new territorial model replaced imperial or tribal associations that proved to be weak and spiritually hollow. The weakness of Middle Eastern tribes against the power of great (Roman or Persian) empires was redeemed by religious ideas that provided a new self-esteem and involved embracing such territorial orders as empire or caliphate (Chapter 1). The Reformation similarly suggested that God would favour the monarch or state as the embodiment of 'good governance' (Chapter 3). This was a shift away from the prevalent conception of *closure* (divine and earthly power concentrated in one centre) – a shift that provoked this moral correction. Nationalism redressed the 19th century shift in *governance* that integrated different classes into a common system of industry and education by making them members of the same (national) family (Chapter 4). Moral justifications help us to see awkward but inevitable conditions as inherently good. This does not mean merely reconciling oneself with a bitter truth; morals may actually open new opportunities and introduce new satisfactions. The late-modern process of globalization, with all its indigestible manifestations, seems to demand a similar moral reform that reconciles us with the new turbulence in *closure,* although the challenges in terms of scale are bigger than ever. Not surprisingly, we see people falling back on the familiar moral strategies of religion and nationalism. However, the current re-territorialization concerns a shift in *closure* that is different from what occurred at the end of the Middle Ages (we are currently territorially unbundling, not bundling). It also differs from the type of *governance* that unleashed 19th century nationalism (at the time incorporation of people in a system of production and education) whereas governance now is obsessed with the challenge of controlling borderless resources, corruption and other volatile powers).

The most conspicuous moral defence against globalization is **(neo)-nationalism**. It conflates the threat of losing the race for economic power with 'losing our country', which supposes both a hostile external world and the penetration of the national space by 'unaccountable' foreign elements (persons and cultural products). It is ironic that the argument that we must retrieve our national cultural heritage is often made by people who never seemed devoted to the places or people that define the national culture in the first place: museums, artists, or historians. And the one who would reject such people and events as 'elite' should also acknowledge that so called 'popular-culture' is either healthier and thriving under the conditions of globalization or in decline because our tastes (for dancing the Quadrille or Charleston) have just moved on. As a moral imperative, nationalism at least inspires solidarity within national borders. However, examples of a greater willingness to redistribute national wealth under a nationalist regime seem pitifully absent. What people often mean with 'losing our country' is a mix of domestic social changes, appearance of unknown faces on TV and the supposed limits on freedom imposed by international agreements. It is significant that UK citizens who voted for 'Brexit' favoured the death penalty and corporal punishment in schools at a rate almost three times higher than those voting 'Remain'.[9] We all suffer from nostalgia in moments of weakness, but it is wise to distinguish between social change (do we miss slavery?) and the urge to make all nations equal.

A more realistic threat that stimulates a movement towards nationalism is migration. Unfortunately, there are parts of the world where life is so precarious that people risk everything to illegally enter countries where a standard of human rights means that one will be fed and will not be shot: a

---

[9] YouGov survey 2017, see http://www.huffingtonpost.co.uk/entry/leave-voters-brexit-day_uk_58db873be4b0cb23e65ccbd2 (accessed 23/2/2018).

step forward for a lot of people. Since sending migrants back is fraught with practical and political difficulties, the receiving state is confronted with the choice between allowing them to enter or lowering the standards of human rights. Abolishing all border control – in the case of the EU the outside border – would create many problems of governance. The conclusion of many nationalists is to ignore the moral standards, a policy that may have epidemic consequences on the world order and eventually strike back on the acting state. The only way out is concerted international action, which is of course an infringement on the worshipped principle of national independence.

The manifestation of nationalism that is currently receiving the most attention is **populism**. Populism flourishes on a number of resentments: inequality of wealth, invasion by hostile elements in the familiar culture, 'elites' that are indifferent to the common man and the failure of democracy as a truly representative system. The most characteristic strategy is the embracing of a strong leader (person or movement) who directly communicates with the people (social media). Other common actions of populist leaders/movements are: ignoring parliament, courts or official media, the subsequent undermining of neutral judicial process, creation of internal and external enemies and holding a highly opportunistic or capricious stance in international relations. As an analysis of voting for populist parties since 1900 has shown, we currently see the highest level of populism since the 1930s, although the ideology of today's populism is less extreme.[10] This analysis included cases like Trump, UKIP in the UK, AfD in Germany, National Front in France, Podemos in Spain and the Five Star Movement in Italy. There are other populist leaders like Orban in Hungary, Modi in India, Duterte in the Philippines and Erdoğan in Turkey. None of these leaders or movements offers a sustainable moral strategy that incorporates globalization partially or wholly.

Religion is a moral strategy that offers a less tormenting answer to globalization than the nationalist strategy. While the success of certain Christian movements, like the Pentecostal Church in Latin America with its glorification of self-realization, is also a consequence of the new social mobility induced by globalization[11]. However, the most obvious religious answer is the worldwide rise in Islamic assertiveness. For centuries, Islam was the silent

---

[10] *Bridgewater Daily Observations*, 2017 (March 22), 'Populism: The Phenomenon'. https://www.bridgewater.com/resources/bwam032217.pdf
[11] Yannick Fer, 2007, 'Pentecôtisme et modernité urbaine: entre déterritorialisation des identités et réinvestissement symbolique de l'espace urbain'.

culture of tribal communities, recognisable in the architectural landscape and in daily prayer rituals but not in a strongly assertive emphasis on rules of life or dress codes. Such customs belonged to the specific tribal culture rather than to religion. This all changed when Western values in the shape of consumerism and secularism entered the Muslim world (for example in Egypt after it became a British colony in 1887). This created a fertile soil for movements like the Muslim Brotherhood (founded in 1928), which proclaimed a return to the strict principles of the Quran both in daily life and state practice.

A second wave of 'fundamentalism' was inspired by the labour-migration of large groups of Muslims to the West, where they not only encountered other dominant values about politics and family life but also other migrants with different customs and varieties of Islam.[12] It created a search for a unified Islam that would be recognisable and verifiable by lifestyles and looks. The veil and the sharia entered local Muslim communities where before they had never played a prominent role. This is clear evidence of the human need to identify with a strong entity, whether it is a state, nation, or transnational religious community. Even in a state like Turkey where politics has been committed to the westernized secular state ideal for most of the 20th century, part of the population gradually embraced Muslim identity as a political guiding principle in the more recent period. This was also encouraged by political entrepreneurs, like president Erdoğan, with the aim of stifling opposition and bracing oneself against claims from the European Union.

Obviously, adopting a religious identity can support nationalist and authoritarian aims, but this obfuscates the entirely autonomous role of religion as a moral reaction to globalization in its neoliberal shape. As the testimonies of Western converts to Islam or young Muslims drifting to the Islamic State have recently divulged, many of them experienced a lack of meaning in the kind of life that is entirely devoted to consumption and improving one's material position, the more so if they were themselves failing in the pursuit of such ideals. This is not a flat-out rejection of globalization, but an attempt to replace its dominant economic drive with more idealistic aims that usually entail a global vision. And the lure of a secluded paradise on Earth – a Caliphate – is never wasted on those who look at world history with millennialist eyes!

---

[12] Olivier Roy, 2004, *Globalized Islam. The Search for a New Ummah*.

There is a philosophical counterpart to the religious criticism of globalization (particularly in its Islamic shape) that accuses globalization of 'westernization'. The focus here is on the dominant Western message of individualization and its deconstruction of all cultural values simply as ways to wield power and deceive people (in the spirit of Michel Foucault). Actually, as its critics say, this (postmodern) philosophy ends in an empty embrace of technological progress, in which machines rule people in the absence of social goals that are beyond suspicion. It is the type of criticism that we also meet among conservative Russian thinkers like Aleksander Dugin (chapter 6) and predecessors like Alexander Herzen or Ivan Ilyin. In line with the Eurasianists, some conservative thinkers are now looking to classic Chinese philosophy as a recipe for reconciling technological progress with a non-individualist conception of society.[13] From a Western point of view, the licence of illiberal practices that this body of thought requires will be unacceptable. However, in its emphasis on social links there is resonance with the following moral adaptation to globalization.

From global street to global stake

One of the ways to incorporate globalization in daily life is the practice of **'glocalization'**. In this case, glocalization denotes the policy horizon expansion of main urban centres and sub-national regions across national and continental boundaries, breaking the norm of hierarchical government in which the national state is the sole gatekeeper of political dealings with the world outside.[14] Further, it represents the positive experience of globality in the local presence of global products, ethnic diversity and cultural innovation. This shift logically resulted from the fact that an urban centre's global economic network has become much more important for its wealth than the national network.[15] However, it cannot be denied that the urban scene also represents the most tenacious forms of ghettoization, gated communities and race riots. Nevertheless, the urban setting is also a vehicle of empowerment because it makes protests particularly noticeable. One of the mechanisms behind the

---

[13] V.V.Maliavin, *Globalization and the moral issue.* http://tkugloba.tku.edu.tw/english/doc-e/globalization.htm (accessed 5/8/2017)
[14] Erik Swyngedouw, 2004, 'Globalization or "glocalisation"? Networks, territories and rescaling'.
[15] Peter Taylor and Ben Derudder, 2016, *World City Network: A Global Urban Analysis*.

'global street' (as Saskia Sassen has called the phenomenon[16]) is its epidemic character, which we encountered above in the mushrooming of 'Occupy' movements. As the recent history of these protests shows, the influence of such movements does not lie in the real exercise of power, but rather in their indirect effects; in Sassen's terms, these effects '(make) powerlessness more complex'. This in turn makes these groups more politically visible and gives city-dwellers the chance to be recognized as a collective claimant. In Berna Turan's study about the local growth of alliances between groups with different lifestyles in neighbourhoods of Istanbul and Berlin, one Istanbul resident protesting the interference of the state (police) in their neighbourhood tells it very succinctly: 'The apartment is rental but the neighbourhood is ours'[17]. This new feeling of collective control that unites people with different life styles and values – representative of global diversity – may be one of the important moral contributions to the public acceptance of globalization.

None of these moral stances entails a straightforward recipe for solving the economic problems that are attributed to globalization, such as inequality or lagging growth. But they may at least alleviate feelings of alienation, and could possibly raise the readiness to implement solidarity and redistribution. By acting as a shop window for worldwide products and ideas, *glocalization* may also stimulate the propensity for innovation and education. Whether one aspires to resource redistribution or universal education, a favourable human disposition is not sufficient to drive actions in the desired direction. Some type of government organization and implementation is indispensable. Here we arrive at the crucial question: do we see a new territoriality in the making, one which has rejected the national competition (governance-identity) paradigm of the past two centuries in order to tackle new uncertainties and exploit new surplus values? Benjamin Barber's vision in *If Mayors Ruled the World*[18] does not anticipate the abolishment of states, but rather a drastic reshuffling of authority and resources between the state and the local (city) level. The underlying idea is that our most pressing problems are global, but states are unable or unwilling to solve them because their (passive) dimension, closure/sovereignty, prevents state-action. Cities, on the contrary have a natural propensity to interact across boundaries because that is their life line. As Barber remarks, 'It is a most remarkable political conundrum that the unique

---

[16] Saskia Sassen, 2010, 'The Global Street: making the political'.
[17] Berna Turam, 2015, *Gaining Freedoms. Claiming Space in Istanbul and Berlin,* p. 178.
[18] Benjamin Barber, 2013, *If Mayors Ruled the World: Dysfunctional Nations, Rising Cities.*

power held by sovereign states actually disempowers them from cross-border cooperation while the corresponding powerlessness of cities facilitates such cooperation'[19] Cities struggle directly with the consequences of climate change, illegal immigrants, terrorism, and social inequality. They are compelled to find practical solutions, which can in turn be shared with other global cities, either in the shape of discussing policy results or in intelligence sharing about threats like terrorism. The success of American cities in confronting the causes and consequences of climate change, a glaring contrast with national politics, is an example. The issue of gun control, promoted by mayors but resisted at the federal or state level, has revealed a similar opposition.

The European Union holds a long-standing commitment to diminish the negative effects of state boundaries, obviously in terms of enhancing mutual understanding between states but also in eliminating the local disruptive effects of state boundaries. The so-called 'territorial cohesion' policy[20] demands attention for the key roles of cities as motors of development, harmoniously connected with their rural surroundings in a polycentric pattern across Europe. In addition, this policy promotes interregional connections in the sphere of transport, ecology, and culture. The main thrust of the program is to exorcize the centre-periphery ghosts of the past when (state-)boundary areas inevitably suffered from lack of national attention, poverty, and weak infrastructure. Cross-border cooperation and EU funds (INTERREG program) were a key to achieving regional exchange, fostering learning about other policy experiences and organizing allocation of resources for technical innovation, sustainable development, cultural manifestations, higher service levels and so on. A border town hospital that serves patients of two countries (in the Pyrenees town of Puigcerda) is an example. In 2016, the European Council for General Affairs adopted an Urban Agenda for the EU. This agenda upgraded the strategic role of cities (in line with territorial cohesion policy) to the responsibility of playing a part in solving 'the many contemporary social, political, economic and environmental problems the states and

---

[19] Ibid., p. 139.
[20] 'Territorial state and perspectives of the European Union', Document prepared for the Informal Ministerial Meeting on Regional Policy and Territorial Cohesion, 20/21 May 2005, Luxembourg
http://www.eu2005.lu/en/actualites/documents_travail/2005/05/20regio/Min_DOC_1_fin.pdf

the EU seem unable to tackle'.[21] Since local authorities implement a large share of the EU regulations and policies, it seemed wise to involve them early in the decision making process. It is expected that this form of multi-level governance may help to introduce common policies that, as previous events showed, might be blocked by unwilling member states in top decision-making at the EU level. Here Benjamin Barber's vision has become fully implemented.

Glocalization addresses the practical merits of shifting a state-dominated territorial paradigm to a new mix of global and local, but how does it involve a new moral outlook and – a related question – how does it gain broad support among people? One advantage of the global city is that it already offers a mix of foreign elements in a familiar setting, while at the same time retaining opportunities to withdraw into a safe social niche. There is nothing wrong with the need for a familiar social environment, but curiosity is another basic human drive. The opportunity to learn something about other cultural habits and aesthetic norms is usually seized upon if the offer is non-totalitarian or non-threatening. For many workers, whether the carpenter, shopkeeper or accountant, cultural diversity may be a source of inspiration and a possibility for jobs in a wider, even international, network. However, globalization, in connection with the instruments of the information age, has more advantages to offer, not only for urban people but also for those who lead traditional and rural lives. From the fruit culture in Tambogrande to the threatened trees in residential quarters of Delhi, locals may count on a wider, even global, public that supports ideals of sustainability and a more fulfilling way of life. This global interest is not merely a noncommittal expression of taste but results from the conviction that any decision in the world may eventually determine one's own chances of survival. One has a stake in other life worlds. This is completely different from the nationalist view of the world. Compare it with two strategies to avoid traffic accidents. People are naturally egocentric, inclined to look at the world from their own perspective. But as any road user knows, it pays – on penalty of an accident – to put oneself in the position of another traffic participant. Nationalism wants to avoid collisions by drastically separating traffic streams. Glocalism pursues the same goal by raising awareness, an approach that better fits the 'music of our time': the information age. When nationalist regulation fails, collisions are awful. When glo-

---

[21] Virginie Mamadouh, 2018, 'The city, the (Member) state, and the European Union'.

calism fails, they are limited in scope.

None of these moral considerations guarantees that there will be a smooth transition to a new territorial order. Nationalism evolved amidst the ruins of Napoleonic wars, and the early-modern state, though considered sacred, had to cope with a period of religious wars. As chapter 6 (Dying States) suggests, there are three possible reactions to the late-modern era of globalization: imperialism, supranationalism and glocalism. Imperialism, with a strong authoritarian system and military threat to other states, may be sustainable when it improves the material situation of a population (like in China) and creates a strong ideology aided by the manipulation of information. But in the long run it cannot outdo the diversity of internal interests and the message of external voices (see Chapter 5: Private legible organizations). Supranationalism, though plagued by uncompromising secessionist drives, is something we cannot do without, as evidenced by the failure of external interference with the Middle East or the threat of climate change. Let's hope that these considerations, and soft approaches like glocalism, may prevent the disasters that the current refuge within national borders has in store.

APPENDIX
The argument in brief

Chapter 1 Territorial Shock: an introduction

Territorial shocks reflect a conflict between an established territorial practice (order) and political changes which demand a new moral confirmation. A territorial order is defined by the implementation of three dimensions: *closure* (who and what is subordinated to an authority), *governance* (resources developed by the authority) and *identity* (the meaning of the territorial order). Globalization is a contemporary source of territorial shock. This book distinguishes between two previous shifts in the evolution of the Western (European) territorial state that aroused reactions fitting the idea of territorial shock: the transformation from an imperial system to a collection of sovereign states at the end of the Middle Ages and the rise of the infrastructural nation-state since the early 19th century.

Chapter 2 Barbarians at the gates: the classic Empires

Closure: the classic empire did not recognize legitimate equals, only various tribes (or kingdoms) that had to be pacified or deterred. Consequently, the Empire's armies regularly crossed the border zone.
Governance: this consisted of numerous intermediaries that were supposed to implement top-down control. Integration depended on bureaucratic discipline (imperial education and awarding of honours), but corruption was difficult to detect.
Identity: the empire as direct extension of heaven. Classified as *passive* because there was no role for other identities (like ethnic distinctions as justification for territorial autonomy).

Chapter 3 A New Jerusalem: the birth of the territorial state

The dissolution of the imperial ideal in the centuries after 1300 made *closure* a hot issue (How to delimit the new kingdoms, particularly if they had liberated themselves? What to do with dispersed territories of a royal dynasty? etc.). Moreover the passive conception of the emperor (or pope) as representative of God became an active *identity* question about the legitimacy of territorial sovereigns. This territorial shock was partly resolved by the Reformation and in other cases by a holy conception of the state or its special guidance by God.

Governance in the hands of the state was dominated by taxation, but many practical tasks continued to be in the hands of local caretakers, nobility or dignitaries. Governance is therefore characterized as *passive*.

Chapter 4 The vertigo of public space (High-Modern territoriality)

The arrival of the 19th century saw the rise of a central state that was bent on making the territory productive by infrastructural control and collecting information (*governance* active). General education in a national language emphasized the mutual dependency of people that before had been completely indifferent to each other. Participating in the same enterprise with 'others' was a source of territorial shock, but was alleviated by the moral imperative of sharing national 'genes' (*identity* active). The idea of a fixed state system (substantiated by a surge in the number of multilateral treaties) made *closure* less enigmatic than in the preceding period (passive).

Chapter 5 Can the centre hold? Territory in the Age of Late-Modernity

The current phase of globalization is characterized by attacks on the independent authority of states from different sides: strong non-state entities (individuals, enterprises, NGOs), failed states accommodating forces hostile to the international order, information systems that directly address the citizen (social media), etc. This turns *closure* into an active and disturbing issue, as demonstrated by attacks on multilateral treaties, retro-nationalism or the erection of walls. Doubts about the democratic system also keep *governance* in an active state. These are crude responses to territorial shock that neither eliminate the new fragmentation of space (porosity of boundaries) nor offer a moral imperative for the new global system.

Chapter 6 Dying states: prelude to re-territorialization?

The assumption of this chapter is as follows: when states have to start from a clean slate, a more radical accommodation to a new world order is more feasible than a prolonged struggle that involves opposing interest groups and political voices. Here we look at France after the Second World War, Russia after the collapse of the Soviet Union and Somalia after a period of collapse that has lasted more than 25 years. France's incorporation into the European Community was a truly qualitative jump. Russia's transformation under Putin is wrapped up in the guise of classic imperialism. Somalia's destiny is still unclear but shows signs of a glocal or (transnational in Somaliland) incorporation into the world under the influence of an active diaspora. These three cases seem to represent the three choices for the world in a global era.

Chapter 7 Globalization and its detractors

Criticism of globalization started with fear of a 'race to the bottom', which would downgrade values like decent working conditions and quality of the environment. A few years later it was the instability of international finance and increasing inequality. In the EU, the institution of a common currency seemed to enhance dependency of certain members. In the Global South the intrusion of enterprises damaging the environment, or the imposition of international norms has incited resistance. None of these is an inevitable consequence of globalization that would necessitate a return to a world of hard spatial containers. On the contrary, international movements may champion local values, and global norms about sustainability may be a local challenge. This is the glocal solution to the perceived danger of globalization.

LITERATURE

Abdulghani, Jasim M., 1984, *Iraq and Iran: The Years of Crisis*. Baltimore : Johns Hopkins University Press.

Agnew, John, 1998, *Geopolitics: Re-Visioning World Politics*. London: Routledge.

Ahmed, Akbar, 2013, *The Thistle and the Drone. How America's War on Terror Became a Global War on Tribal Islam*. Washington : The Brookings Institution.

Alexievich, Svetlana, 2016, *Secondhand Time: The Last of the Soviets*. New York : Random House.

Al-Rawi, Ahmed, 2016, 'Video games, terrorism, and ISIS's Jihad 3.0'. *Terrorism and Political Violence*. DOI: 10.1080/09546553.2016.1207633

Anderson, Benedict, 1985, *Imagined Communities: Reflections on the Origin and Spread of Nationalism*. London: Verso.

Axtmann, Ronald, 2003, 'State formation and supranationalism in Europe: The Case of the Holy Roman Empire of the German Nation'. In: Mabel Berezin and Martin Schain, *Europe Without Borders: Remapping Territory, Citizenship and Identity in a Transnational Age*, pp. 118-139.

Baldwin, Richard, 2016, *The great Convergence. Information, Technology and the New Globalization*. Cambridge (Mass.) : The Belknap Press of Harvard University Press.

Barber, Benjamin, 2013, *If Mayors Ruled the World: Dysfunctional Nations, Rising Cities*. New Haven : Yale University Press.

Barfield, Thomas J., 1989, *The Perilous Frontier. Nomadic Empires and China*. London : Basil Blackwell.

Bartlett, Robert and Angus Mackay, *Medieval Frontier Societies,* 1989, Oxford : Clarendon Press.

Bassin, Mark, 1999, *Imperial Visions: Nationalist Imagination and Geopolitical Expansion in the Russian Far East*. Cambridge : Cambridge University Press.

Bassin, Mark, 2015, 'Lev Gumilev and the European New Right', *Nationalities Papers* 43(6), 840-865. DOI: 10.1080/00905992.2015. 1057560

Baxter, Joan, 2010, 'Great African land grab'. *Le Monde diplomatique* (English ed.), april 2010.

Bayart, Jean-Francois, 2004, *Global Subjects. A Political Critique of Globalization*. Cambridge : Polity Press.

Bayly, Christopher A. and Eugenio F. Biagini, 2008, *Giuseppe Mazzini and the Globalisation of Democratic Nationalism 1830-1920*. Oxford: Oxford University Press.

Beck, Ulrich, 1986, *Risikogesellschaft. Auf dem Weg in eine andere Moderne*. Frankfurt a. M. : Suhrkamp.

Beissinger, Mark R., 2002, *Nationalist Mobilization and the Collapse of the Soviet State*. Cambridge : Cambridge University Press.

Bell, David A., 2001, *The Cult of the Nation in France: Inventing Nationalism, 1680-1800*. Cambridge (Mass.) : Harvard University Press.
Bentham, Jeremy, 1843, *The Works of Jeremy Bentham published under the superintendence of his executor, John Bowring, vol. ix*. Edinburgh : William Tait.
Berezin, Mabel and Martin Schain, 2003, *Europe Without Borders: Remapping Territory, Citizenship and Identity in a Transnational Age*. Baltimore : The Johns Hopkins University Press.
Berg, Eiki and Wim van Meurs, 2002, 'Borders and orders in Europe: limits of nation- and state-building in Estonia, Macedonia and Moldova'. *Journal of Communist Studies and Transition Politics* 18(4), pp. 1-74.
Berger, Suzanne et. al., 2005, *How We Compete. What Companies Are Doing to Make it in Today's Global Economy*. New York : Currency – Doubleday.
Berman, Marshall, 1982, *All that is solid melts into air. The experience of modernity*. New York : Simon and Schuster.
Bira, Sh., 2004, 'Mongolian Tenggerism and modern globalism'. *Journal of the Royal Asiatic Society* 14(1), pp. 3-11.
Black, Jeremy, 2001, *Eighteenth-Century Britain, 1688-1783*. Palgrave : Houndmills.
Blanning, T.C.W., 2002, *The Culture of Power and the Power of Culture: Old Regime Europe 1660-1789*. Oxford : Oxford University Press.
Blom, J.C.H. and E. Lamberts, 1999, *History of the Low Countries*. New York : Berghahn Books.
Bordua, David J. (ed.), 1967, *The Police. Six Sociological Essays*. New York : John Wiley & Sons.
Boyle, Nicholas and Martin Swales, 1986, *Realism in European Literature*. Cambridge : Cambridge University Press.
Bracewell, Wendy, Tamara Dragadze and Anthony Smith, 1993, *Pre-modern and Modern National Identity in Russia and Eastern Europe*. Yverdon : Gordon and Breach.
Bridgewater, 2017, 'Populism: The Phenomenon'. *Bridgewater Daily Observations* 2017 (March 22), https://www.bridgewater.com/resources/bwam032217.pdf
Browder, William, 2009, *Hermitage Capital, the Russian State and the Case of Sergei Magnitsky*. Chatham House REP Round Table Summary 15/12/2009.
Brown, Peter, 2013, *The Rise of Western Christendom: Triumph and Diversity, A.D. 200-1000*. Chichester : Wiley-Blackwell.
Brubaker, Rogers, 1994, 'Nationhood and the national question in the Soviet Union and Post-Soviet Eurasia: an institutionalist account', *Theory and Society*, vol. 23(1), pp. 47-78.
Bruneteau, Bernard, 2003, *'L'Europe Nouvelle' de Hitler. Une Illusion des Intellectuels de la France de Vichy*. Monaco : Éditions du Rocher.

Budde, Gunilla, Sebastian Conrad and Oliver Janz, 2000, *Transnationale Geschichte, Themen, Tendenzen und Theorien*. Göttingen : Vandenhoeck & Ruprecht.
Bull, Hedley N., 1977, *The Anarchical Society: A Study of Order in World Politics*. London : Macmillan.
Bunnelll, Tim, Hamzah B. Muzaini and James D. Sidaway, 2006, 'Global city frontiers: Singapore's hinterland and the contested socio-political geographies of Bintan, Indonesia'. *GaWC Research Bulletin* 182 (http://www.lboro.ac.uk/gawc/rb/rb182.html).
Burgess, Glenn ed., 1999, *The New British History. Founding a Modern State 1603-1715*. London : I.B. Tauris.
Burgess, Michael and Hans Vollaard, 2006, *State Territoriality and European Integration*. London : Routledge.
Busch, Werner, 2008, *Caspar David Friedrich: Ästhetik und Religion*. München : Verlag C.H. Beck.
Carré, John le, 2013, *A Delicate Truth*. London : Viking (Penguin Books).
Callaghy, Thomas, Ronald Kassimir and Robert Latham, 2001, *Intervention & Transnationalism in Africa. Global-Local Networks of Power*. Cambridge : Cambridge University Press.
Castells, Manuel, 1998, *End of Millennium (The Information Age: Economy, Society and Culture III)*. Oxford : Blackwell.
Clark, Christopher, 2007, *Iron Kingdom. The Rise and Downfall of Prussia 1600-1947*. London, Penguin.
Clément, Caty, 2004, *State Collapse: A Common Causal Pattern? A Comparative Analysis of Lebanon, Somalia and the Former-Yugoslavia*. Louvain : Catholic University of Louvain (PhD thesis).
Colley, Linda, 1992, *Britons. Forging the nation 1707-1837*. London : Pimlico.
Collins, James B., 1997, 'State building in Early-Modern Europe: the case of France'. *Modern Asian Studies* Vol. 31(3), pp. 603-633.
Crone, Patricia, 1987, *Meccan Trade and the Rise of Islam*. Princeton : Princeton University Press.
Crossan, John Dominic and Jonathan L. Reed, 2005, *In Search of Paul. How Jesus's Apostle Opposed Rome's Empire with God's Kingdom*. New York : HarperCollins.
Dahl, Tove Stang, 1977, 'State intervention and social control in nineteenth-century Europe'. *Contemporary Crises* 1(2), 163-187.
Darling, Linda T., 'Reformulating the *Gazi* narrative : When was the Ottoman state a *Gazi* state?', *Turcica* 43 (2011), pp. 13-53.
Davies, Thomas, 2013, *NGOs: A New History of Transnational Civil Society*. London : Hurst & Company.
Delaney, David, 2005, *Territory: A Short Introduction*. Blackwell : Malden(MA) etc.

Deleuze, Gilles. and Guattari, F., 1980, *Mille Plateaux : Capitalisme et Schizophrénie*, Paris, Les Éditions de Minuit.
Denemark, Robert A.and Matthew J. Hoffmann, 2008, *Global Diplomacy and World System History: A Network analysis of the Multilateral Treaty System over 400 Years*. Paper presented at the annual meeting of the ISA's 49th annual convention, *Bridging Multiple Divides*, San Francisco, CA, Mar 26, 2008.
Desprairies, Cécile, 2013, *L'Héritage de Vichy: Ces 100 mesures toujours en vigueur*. Paris : Armand Colin.
Deursen, A. Th. Van, 1999, 'The Dutch Republic, 1588-1780'. In J.C.H. Blom & E. Lamberts, *History of the Low Countries*, pp. 143-220.
Dietz, Ton and Dick Foeken, 2001, 'The crumbling of the African state system'. In: Gertjan Dijkink and Hans Knippenberg (Eds.), *The Territorial Factor*, pp. 177-200.
Dijkink, Gertjan, 1996, *National Identity and Geopolitical Visions. Maps of Pride & Pain*. London : Routledge.
Dijkink, Gertjan and Hans Knippenberg (eds.), 2001, *The Territorial Factor: Political Geography in a Globalizing World*. Amsterdam : Vossiuspers University of Amsterdam.
Dijkink, Gertjan and Inge van der Welle, 2009, 'Diaspora and sovereignty: three cases of public alarm in The Netherlands'. *Tijdschrift voor Economische en Sociale Geografie/ Journal of Economic and Social Geography* 100(5), pp. 623-634.
Dijkink, Gertjan, 2006, 'When religion and geopolitics fuse: a historical perspective'. *Geopolitics* 11(2), pp. 192-208.
Dijkink, Gertjan, 2008, 'Nationalism and geopolitics'. In: Guntram H. Herb and David H. Kaplan, *Nations and Nationalism*, pp. 458-470.
Dijkink, Gertjan, 2010, 'Territorial Shock: toward a theory of change'. *L'Espace Politique* 12(3). https://journals.openedition.org/espacepolitique/1781
Drake, David, 2002, *Intellectuals and Politics in Post-War France*. Houndmills : Palgrave.
Drescher, S., D. Sabean, A. Sharlin (eds.), 1982, *Political Symbolism in Modern Europe*, Transaction Books : New Brunswick.
Dreyfus, François-Georges, 1990, *Histoire de Vichy*. Paris : Perrin.
Duby, Georges, 1991, *France in the Middle Ages*. Cambridge : Cambridge University Press.
Duffy, Rosaleen, 2005, 'Global environmental governance and the challenge of shadow states: the impact of illicit sapphire mining in Madagascar', *Development and Change* 36(5), pp. 825-843.
Dugin, Alexander, 2012 (2009), *The Fourth Political Theory*. London : Arktos.
Dunford, Michael Frederick and Grigoris Kafkalas (Eds.), 1992, *Cities and Regions in the New Europe*. London : Belhaven Press.

Dyson Stephen L, 1985, *The Creation of the Roman Frontier*. Princeton : Princeton University Press.
Dyson, Kenneth, 1980, *The State Tradition in Western Europe. A Study of an Idea and Institution*. Oxford : Martin Robertson.
Elden, Stuart, 2013, *The Birth of Territory*. Chicago : University of Chicago Press.
Engvall Johan, 2015, 'The state as investment market: a framework for interpreting the Post-Soviet state in Eurasia'. *Governance: an International Journal of Policy, Administration and Institutions* 28(1), pp. 25-40.
Engvall, Johan, 2012, *Against the Grain. How Georgia fought Corruption and what it Means*. Washington : Central Asia-Caucasus Institute & Silk Road Studies Program (www.silkroadstudies.org).
Fairbank, John K. (ed.), 1968, *The Chinese World Order. Traditional China's Foreign Relations*. Cambridge (Mass.) : Harvard University Press.
Fazal, Tanisha, 2007, *State Death. The Politics and Geography of Conquest, Occupation and Annexation*. Princeton N.J. : Princeton University Press.
Fer, Yannick, 2007, 'Pentecôtisme et modernité urbaine: entre déterritorialisation des identités et réinvestissement symbolique de l'espace urbain'. *Social Compass* 54(2), pp. 201-210.
Ferrante, Elena, 2015, *The Story of the Lost Child* (Neapolitan Novels Book 4). New York : Europa Editions.
Fletcher, Joseph F., 1968, 'China and Central Asia, 1368-1884'. In:. John K. Fairbank, *The Chinese World* Order, pp. 206-224.
Friedeburg, Robert von, 2005, 'The making of patriots: love of fatherland and negotiating monarchy in seventeenth-century Germany'. *The Journal of Modern History*, 77(4), pp. 881-916.
Gachechiladze, Revaz, 1995, *The New Georgia: Space, Society, Politics*. London : UCL Press.
Galster Ingrid, 2001, *Sartre, Vichy et les Intellectuels*. Paris : l' Harmattan.
Galster, Ingrid, 2001, *La naissance du Phénomène Sartre. Raisons d'un succès 1938-1945*. Paris : Ed. du Seuil.
Gamlen, Alan, 2006, *Diaspora engagement policies: what are they, and what kinds of states use them?* Centre on Migration, Policy and Society (COMPAS), University of Oxford. Working Paper No. 32.
Gellner, Ernest, 1983, *Nations and Nationalism*. Oxford : Blackwell.
Gikandi, Simon, 2012, 'The politics and poetics of national formation. Recent African writing and Maps'. In: Derek Wright ed., *Emerging Perspectives on Nuruddin Farah*, pp. 449-467.
Githongo, John, 2005, *Report on the findings of graft in the Government of Kenya, to H.E. President Mwai Kibaki* (summary document), 22 November 2005.
Goffmann, Daniel, 2002, *The Ottoman Empire and Early Modern Europe*. Cambridge : Cambridge University Press.

Goldstein, Robert J. (ed.), 2000, *The War for the Public Mind. Censorship in Nineteenth-Century Europe.* Westport (Conn.) : Praeger.
Grave, Johannes, 2001, *Caspar David Friedrich und die Theorie des Erhabenen.* Weimar : Verlag und Datenbank für Geisteswissenschaften.
Greenfeld, Liah, 1993, *Nationalism. Five Roads to Modernity.* Cambridge(Mass.) : Harvard University Press.
Grenz, Stanley J. and R.E. Olson, 1992, *20th Century Theology: God & the World in a Transitional Age.* Downers Grove (Ill.) : InterVarsity Press.
Groenveld, Simon, 1990, *Evidente factiën in den Staet. Sociaal-politieke Verhoudingen in de 17e-eeuwse Republiek der Verenigde Nederlanden* [Revealed Factions in the State. Social-political Relations in the 17th century Republic of the United Netherlands]. Hilversum : Verloren.
Gross, Bertram M, 1980, *Friendly Fascism: The New Face of Power in America.* New York : Evans and Company Inc.
Grundy-Warr, Carl, Karen Peachey and Martin Perry, 1999, 'Fragmented integration in the Singapore-Indonesian border zone: Southeast Asia's "Growth Triangle" against the global economy'. *International Journal of Urban and Regional Research* 23(3), pp. 304-328.
Haarstad, Håvard and Arnt Fløysand, 2007, 'Globalization and the power of rescaled narratives: A case of opposition to mining in Tambogrande, Peru'. *Political Geography* 26(3), pp. 289-308.
Habermas, Jürgen, 1989 (1962), *The Structural Transformation of the Public Sphere: An Inquiry into a Category of Bourgeois Society.* Cambridge Massachusetts : The MIT Press.
Hale, John K., 1996, 'England as Israel in Milton's writings'. *Early Modern Literary Studies* , vol. 2.2 (http://purl.oclc.org/emls/02-2/halemil2.html).
Hall, John A. (ed.), 1998, *The State of the Nation.* Cambridge : Cambridge University Press.
Hammond, Laura et al, 2012, *Cash and Compassion: The Role of the Somali Diaspora in Relief, Development, and Peace-building.* London : Chatham House (Meeting summary),
http://www.chathamhouse.org/sites/default/files/public/Research/-Africa/070312somalidiaspora.pdf).
Harding, Andrew, 2016, *The Mayor of Mogadishu.* New York : St. Martin's Press.
Hardt, Michael and Antonio Negri, 2000, *Empire.* Harvard University Press : Cambridge (Mass.).
Hartmann, K., 1992, *Zwanzig Jahrhunderte Kirchengeschichte. Vom Anfang bis zur Gegenwart.* Lahr : Ernst Kaufmann Verlag.
Hay, John (ed.), 1995, *Boundaries in China.* London : Reaktion Books.
Heffernan, Michael, 2005, 'Geography, empire and National Revolution in Vichy France'. *Political Geography* 24(6), 731-758.

Heijden, H.A.M. van der, 1998, *Oude kaarten der Nederlanden 1548-1794 / Old maps of the Netherlands 1548-1794*. Canaletto/Repro-Holland : Alphen aan den Rijn and Universitaire Pers Leuven.
Helleiner, Eric, 2002, 'Economic nationalism as a challenge to economic liberalism? Lessons from the 19th century'. *International Studies Quarterly* 46(3), pp. 307-329.
Hellman, John, 2001, 'Memory, history and national identity in Vichy France'. *Modern & Contemporary France* vol. 9(1), pp. 37-42.
Henne, Peter S, 2013, 'Is all jihad local? Transnational contention and political violence in Somalia'. In: Emma Leonard and Gilbert Ramsay eds., *Globalizing Somalia*.
Herb, Guntram H. and David H. Kaplan, 2008, *Nations and Nationalism: A Global Historical Overview*, vol. 2. Santa Barbara: ABC-CLIO.
Hermand, Jost, 1982, 'Dashed hopes: on the painting of the wars of liberation'. In: S. Drescher, D. Sabean, A. Sharlin (eds.), *Political Symbolism in Modern Europe*, pp. 216-238.
Heyden, H.A.M. van der, 2001, *Kaart en Kunst van de Zeventien Provincies der Nederlanden* [Map and Art of the Seventeen Provinces of the Netherlands]. Canaletto / Repro-Holland and Universitaire Pers Leuven.
Higgins, Iain Macleod, 1998, 'Defining the Earth's center in a medieval multi-text. Jerusalem in The Book of John Mandeville'. In: Sylvia Tomasch and Sealy Gilles, *Text and Territory*, pp. 29-53.
Hindle, Steve, 2000, *The State and Social Change in Early Modern England, 1550-1640*. Palgrave : Houndmills.
Hirst, Paul, 2001, 'Politics: territorial or non-territorial?', *The Global Site, First Press* https://web.archive.org/web/20010620102350/http://www.theglobal-site.ac.uk:80/press/104hirst.htm
Hooson, D. (ed.), 1995, *Geography and National Identity*. Oxford : Blackwell.
Hoover, Arlie J., 1989, *God, Germany and Britain in the Great War: A Study in Clerical Nationalism*. New York : Praeger.
Imber, Colin, 2002, *The Ottoman Empire 1300-1650. The Structure of Power*. Houndmills : Palgrave Macmillan.
Jackson, Julian, 2001, *France: The dark years 1940-1944*. Oxford: Oxford University Press.
Jackson, Michael, 2005, 'The eighteenth century antecedents of bureaucracy, the Cameralists'. *Management Decision* 43(10), pp. 1293-1303.
Jeffries, Ian, 2004, *The Countries of the Former Soviet Union at the Turn of the Twenty-first Century. The Baltic and European States in Transition*. London: Routledge.
Jelavich, Charles and Barbara, 1977, *The Establishment of the Balkan National States, 1804-1920*. Seattle : University of Washington Press.

Jiménez, Manuel González, 1989, 'Frontier and settlement in the kingdom of Castile (1085-1350)'. In: Robert Bartlett and Angus Mackay, *Medieval Frontier Societies*, pp. 49-74.

Jones, Philip, 1997, *The Italian City-State. From Commune to Signoria*. Oxford : Clarendon Press

Jones, Reece, 2012, *Border Walls: Security and the War on Terror in the United States, India and Israel*. London : Zed Books.

Kaiser, Robert J., 1994, *The geography of nationalism in Russia and the USSR*. Princeton : Princeton University Press.

Kantorowicz, Ernst, 1957, *The King's two Bodies*. Princeton : Princeton University Press.

Kaplan, Robert D., 1994, 'The coming anarchy. How scarcity, crime, overpopulation, tribalism, and disease are rapidly destroying the social fabric of our planet'. *The Atlantic Monthly*, February 1994.

Kaplan, Robert D., 2009, 'The revenge of geography'. *Foreign Policy* 172, May/June, pp. 96-105.

Kappeler, Andreas, 1993, 'Some remarks on Russian national identities (sixteenth to nineteenth centuries)'. In: Wendy Bracewell et al, *Pre-modern and Modern National Identity in Russia and Eastern Europe*, pp. 147-155.

Kar, Dev and Sarah Freitas, 2013, *Russia: illicit financial flows and the role of the underground economy*. Washington : Center for International Policy.

Karateke, Hakan T, 2005, 'Legitimizing the Ottoman sultanate: a framework for historical analysis'. In: H. Karateke and M. Reinkowski (eds.), *Legitimizing the Order: The Ottoman Rethoric of State Power*. Leiden : Brill, pp. 13-52.

Kaufmann, Daniel, Aart Kraay and Massimo Mastruzzi, 2010, *The Worldwide Governance Indicators. Methodological and Analytical Issues*. The World Bank, Policy Research Paper WPS 5430.

Kedourie, Elie, 1994 (1960), *Nationalism*. Oxford: Blackwell.

Kepel, Gilles and Antoine Jardin, 2015, *Terreur dans l'Hexagone. Genèse du Djihad Français*. Paris : Gallimard.

King, Charles and Neil J. Melvin, 1999/2000, 'Diaspora politics: ethnic linkages, foreign policy and security in Eurasia'. *International Security* 24(3), pp. 108-138.

Knippenberg, Hans, 1997, 'Dutch nation-building. A struggle against the water?', *GeoJournal* vol. 43(1), pp. 27-40.

Knoll, Paul, 1989, 'Economic and political institutions on the Polish-German frontier in the Middle Ages: action, reaction, interaction'. In: Robert Bartlett and Angus Mackay, pp. 151-174.

Köhler, Joachim, 1997, *Wagners Hitler. Der Prophet und sein Volstrecker*. München : Karl Blessing Verlag.

Kohn Hans, 1944, *The Idea of Nationalism: A Study in its Origins and Background,* New York : Macmillan.

Kolossov, Vladimir and John O'Loughlin, 1998, 'Pseudo-States as harbingers of a new geopolitics: the example of the Trans-Dniester Moldovan Republic'. In: David Newman ed., *Boundaries, Territory and Postmodernity,* pp. 151-176.

Kong, Lily and Brenda S.A. Yeoh, 2003, *The Politics of Landscapes in Singapore. Construction of 'Nation'.* Syracuse (NY) : Syracuse University Press.

Kossmann, E.H., 2000, *Political thought in the Dutch Republic. Three studies.* Amsterdam : Koninklijke Nederlandse Akademie van Wetenschappen.

Krasner, Stephen D., 1999, *Sovereignty: Organized Hypocrisy.* Princeton N.J.: Princeton University Press.

Kristof, Ladis K. D., 1959, 'The nature of frontiers and boundaries'. *Annals of the Association of American Geographers,* 49(3), 269-282.

Larsen Joseph, 2017, *Georgia-China Relations: The Geopolitics of the Belt and Road.* Tbilisi : Georgian Institute of Politics.

Latham, Robert, Ronald Kassimir and Thomas Callaghy, 2001, 'Introduction: transboundary formations, interventions, order, and authority'. In: Thomas Callaghy et al., *Intervention & Transnationalism in Africa,* pp. 1-20.

Lattimore, Owen, 1959, 'Origins of the Great Wall in China: a frontier concept in theory and practice'. In: Owen Lattimore, 1959, *Studies in Frontier History. Collected Papers 1928-1958.* Paris : Mouton & Co., pp. 97-118.

Leonard, Emma and Gilbert Ramsay eds., 2013, *Globalizing Somalia. Multilateral, International and Transnational Repercussions of Conflict.* New Directions in Terrorism Studies 3. New York : Bloomsbury Academic.

Leupen Piet, 1998, *Keizer in zijn eigen Rijk. De geboorte van de nationale staat.* [Emperor in his own realm. The birth of the national state]. Amsterdam : Wereldbibliotheek.

Leustean, Lucian N., 2018, Eastern Orthodoxy, geopolitics and the 2016 'Holy and Great Synod of the Orthodox Church', *Geopolitics* 21(1), 201-216.

Lijphart, Arend, 1969, 'Consociational Democracy'. *World Politics* 21(2), pp. 207-225.

Loewe, Michael, 2006, *The Government of the Qin and Han Empires 221BCE - 220CE,* Indianapolis : Hackett Publishing Company.

Loriaux, Michel, 2008, *European Union and the deconstruction of the Rhineland frontier.* Cambridge: Cambridge University Press.

Loveman, Mara, 2005, 'The modern state and the primitive accumulation of symbolic power'. *American Journal of Sociology* 110(6), pp. 1651-1683.

Lucini, Barbara, 2014, *Disaster Resilience from a Sociological Perspective: Exploring Three Italian Earthquakes as Models for Disaster Resilience Planning.* Cham etc. : Springer.

Mackinder, Halford, 1904, 'The geographical pivot of history'. *The Geographical Journal* 23(4), pp. 421-444.
Maier, Charles S., 2000, 'Transformations of territoriality, 1600-2000'. In: Gunilla Budde et al., *Transnationale Geschichte, Themen, Tendenzen und Theorien,* pp. 32-55.
Maier, Charles S., 2016, *Once Within Boundaries. Territories of Power, Wealth and Belonging since 1500.* Cambridge (Mass.) : The Belknap Press of Harvard University Press.
Maliavin, V.V., *Globalization and the moral issue.* http://tkugloba.tku.edu.tw/english/doc-e/globalization.htm
Mamadouh, Virginie, 2018, 'The city, the (Member) state, and the European Union'. *Urban Geography* DOI: 10.1090/02723638.2018.1453453.
Mancall, Mark, 1968, 'The Qing tribute system: an interpretive essay'. In : John K. Fairbank (ed.), *The Chinese World Order. Traditional China's Foreign Relations.* Cambridge (Mass.), Harvard University Press, pp. 63-89.
Mann, Michael, 1984, 'The autonomous power of the state: its origins, mechanisms and results'. *Archives Européennes de Sociologie* 25(2), pp. 185-213.
March, James G. and Johan P. Olsen, 1998, 'The institutional dynamics of international political orders'. *International Organization* 52(4), pp. 943-969.
Markusse, Jan, 1997, Power sharing and Consociational democracy in South-Tyrol. *GeoJournal* 43(1), pp. 77-90.
Martin, John and Dennis Romano (eds.), 2000, *Venice reconsidered. The history and civilization of an Italian city-state, 1297-1797.* Baltimore : The Johns Hopkins University Press.
Martines, Lauro, 1980, *Power and Imagination. City-States in Renaissance Italy.* New York : Vintage Books.
McCoy, Charles S. and J. Wayne Baker, 1991, *Fountainhead of Federalism. Heinrich Bullinger and the Covenantal Tradition.* Louisville, Westminster : John Knox Press.
Medina, Leandro and Friedrich Schneider, *Shadow economies around the world. What did we learn over the last 20 years?* IMF Working Paper (WP/18/17) 2018.
Melderis, Hans, 2001, *Raum-Zeit-Mythos. Richard Wagner und die modernen Naturwissenschaften.* Hamburg : Europäische Verlagsanstalt.
Melucci, Alberto, 1996, Challenging Codes: Collective Action in the Information Age. Cambridge : Cambridge University Press.
Menkhaus, Ken, 2004, *Somalia: State Collapse and the Threat of Terrorism.* Oxford : Oxford University Press (Adelphi papers 364).
Merriman, John, 1990, *A History of Modern Europe. Volume I.* London : W.W. Norton & Company.
Milward, Alan, 1992, *The European Rescue of the Nation State.* Berkeley : University of California Press.

Minzner, Carl, 2018, *End of an Era: How China's Authoritarian Revival is Undermining its Rise*. New York : Oxford University Press.
Mohammad, Abdullah A., 2002, *State Collapse and Post-Conflict Development in Africa. The Case of Somalia 1960-2001*. Amsterdam : University of Amsterdam (Ph.D. thesis).
Moïsi, Dominique, 2007, 'The clash of emotions: fear, humiliation, hope and the new world order'. *Foreign Affairs* 86(1), pp. 8-13.
Morgenthau, Hans J., 1962, *Politics in the Twentieth Century. Vol 1 The Decline of Democratic Politics*. Chicago: University of Chicago Press.
Müller, Adam, 1808/09, *Vom Geiste der Gemeinschaft*. Alfred Kröner Verlag : Leipzig (1931).
Needham, Joseph, 1959, *Science and Civilisation in China. Vol. 3: Mathematics and the Sciences of the Heavens and the Earth*. Cambridge, Cambridge University Press.
Newman, David ed., 1998, *Boundaries, Territory and Postmodernity*. London : Frank Cass.
Nimwegen, Olaf van, 2006 'The quest for security: the case of the Dutch Republic'. In: Michael Burgess and Hans Vollaard, *State Territoriality and European Integration*, pp. 17-36.
O'Brien, Conor Cruise, 1998, *God Land. Reflections on Religion and Nationalism*. Cambridge (Mass.) : Harvard University Press.
O'Leary, Brian, 1998, 'Ernest Gellner's diagnoses of nationalism: a critical overview, or, what is living and what is dead in Ernest Gellner's theory of nationalism'. In: John A. Hall (ed.), *The State of the Nation*, pp. 40-88.
O'Loughlin, John, Vladimir Kolossov and Andrei Tchepalyga, 1998, National construction, territorial separatism and post-Soviet Geopolitics: the example of the Transdniester Moldovan Republic. *Post-Soviet Geography and Economics* 39 (6), pp. 332-358.
Ohmae, Kenneth, 1990, *The Borderless World. Power and Strategy in the Global Marketplace*. New York : Harper Collins.
Özkirimli, Umut, 2000, *Theories of Nationalism. A Critical Introduction*. Houndmills : Palgrave.
Paasi, Anssi, 1986, *The Institutionalization of Regions : Theory and Comparative Case Studies*, Joensuu : Joensuunyliopisto.
Pabst, Adrian, 2012, 'The secularism of post-secularity: religion, realism and the revival of grand theory in IR'. *Review of International Studies* 38(05) 995-1017. DOI: 10.1017/S0260210512000447
Pavlowitch, Stevan, K., 1999, *A History of the Balkans 1804-1945*. London : Longman.
Pincus, Debra, 2000, 'Hard times and ducal radiance. Andrea Dandolo and the construction of the ruler in fourteenth-century Venice'. In: John Martin

and Dennis Romano (eds.), *Venice reconsidered. The history and civilization of an Italian city-state, 1297-1797*, pp. 90-136.

Phillips, Sarah G., 2016, 'When less was more : external assistance and the political settlement in Somaliland'. *International Affairs* 92(3), pp. 629-645.

Prunier, Gérard, 2010, 'Le Somaliland, une exception africaine'. *Le Monde diplomatique*, October 2010.

Ramonez, Ignacio, 1996, *Geopolitics of Chaos. Internationalization, Cyberculture & Political Chaos*. New York : Algora Publishing.

Ratzel, Friedrich, 1897, *Politische Geographie oder die Geographie der Staaten, des Verkehres und des Krieges*. München : R.Oldenbourg.

Recchia, Stefano and Nadia Urbinati Eds., 2009, *A Cosmopolitanism of Nations: Giuseppe Mazzini's Writings on Democracy, Nation Building, and International Relations*. Princeton : Princeton University Press.

Reno, William, 2001, 'How sovereignty matters: international markets and the political economy of local politics in weak states'. In: Thomas Callaghy et al., 2002, *Intervention and Transnationalism in Africa*, 2001, pp. 197-215.

Reno, William, 2005, 'The politics of violent opposition in collapsing states'. *Government and Opposition* 40(2), pp. 127-151.

Reynolds, Amy and Barnett, Brooke, 2003, '"America under attack": CNN's verbal and visual framing of September 11'. In: Steven Cherniak, Frankie Y. Bailey, and Michelle Brown, *Media Representations of September 11*. Westport Conn. : Praeger, pp. 85-102.

Rich, Norman, 1974, *Hitler's War Aims. II The Establishment of the New Order*. London : Andre Deutsch.

Rietbergen, Peter J.A.N., 1992, Beeld en zelfbeeld. 'Nederlandse identiteit' in politieke structuur en politieke cultuur tijdens de Republiek [Image and self-image. 'Dutch identity' in political structure and political culture during the Republic]. *Bijdragen en Mededelingen betreffende de Nederlandse Geschiendenis*, vol. 107(4), pp. 635-656.

Ringmar, Erik, 1996, *Identity, Interest and Action. A Cultural Explanation of Sweden's Intervention in the Thirty Years War*. Cambridge : Cambridge University Press.

Roberts, Michael, 1953 (1958), *Gustavus Adolphus. A History of Sweden 1611-1632*. London : Longmans.

Robic, M.-C., 1995, 'National identity in Vidal's Tableau de la Géographie de la France: From Political Geography to Human Geography'. In: D. Hooson (ed.), *Geography and National Identity*. Oxford : Blackwell.

Roitman, Janet, 2001, 'New sovereigns? Regulatory authority in the Chad basin', in Thomas Callaghy et al., 2001, *Intervention & Transnationalism in Africa*, 240-263.

Rorlich, Azade-Ayşe, 1986, *The Volga Tatars. A Profile in National Resilience*. Stanford : Hoover Institution Press.

Rothschild, Emma, 2001, *Economic Sentiments. Adam Smith, Condorcet and the Enlightenment.* Cambridge (Mass.) : Harvard University Press.
Roy, Olivier, 2004, *Globalized Islam. The Search for a New Ummah.* New York : Columbia University Press.
Rubinstein, Jonathan, 1973, *City Police.* New York : Ballantine Books.
Ruggie, John Gerard, 1993, 'Territoriality and beyond: problematizing modernity in international relations', *International Organization* 47(1), pp. 139-174.
Sahlins, Peter, 1989, *Boundaries: The Making of France and Spain in the Pyrenees.* Berkeley : University of California Press.
Samatar, Ahmed I., 1994, *The Somali Challenge. From Catastrophe to Renewal?* Boulder : Lynne Riener.
Sassen, Saskia, 2006, *Territory, Authority, Rights: From Medieval to Global Assemblages.* Princeton: Princeton University Press.
Sassen, Saskia, 2010, 'The Global Street: making the political'. *Globalizations* 8(5), 573-579.
Schama, Simon, 1987, *The Embarrassment of Riches: an Interpretation of Dutch Culture in the Golden Age.* New York : Knopf.
Scott, James C., 1998, *Seeing Like a State. How Certain Schemes to Improve the Human Condition Have Failed.* New Haven : Yale University Press.
Shekhovtsov, Anton, 2009, 'Aleksandr Dugin's Neo-Eurasianism: the New Right à la Russe'. *Religion Compass* 3(4), pp. 697-716.
Shirer, William L., 1970, *The Collapse of the Third Republic.* London : William Heinemann / Secker & Warburg.
Showalter, Daniel, 1998, 'Churches in Context: The Jesus Movement in the Roman World'. In: Michael D. Coogan, *The Oxford History of the Biblical World,* New York : Oxford University Press, pp. 388-419.
Sibley, Chris G and Bulbulia, Joseph, 2012, 'Faith after an earthquake: a longitudinal study of religion and perceived health before and after the 2011 Christchurch New Zealand earthquake'. *PLOS ONE* 7(12): e49648. DOI:10.371/journal.pone.0049648.
Silver, Alan, 1967, 'The demand for order in civil society: a review of some themes in the history of urban crime, police and riot'. In: David J. Bordua (ed.), *The Police. Six Sociological Essays*, pp. 1-24.
Small, Melvyn and J. David Singer, 1970, 'Patterns in international warfare, 1816-1965'. *Annals of the American Academy of Political and Social Science* 391, 145-155.
Smith, Neil, 1996, Spaces of vulnerability: the space of flows and the politics of scale. *Critique of Anthropology,* vol.16(1), pp. 63-77.
Smith, Noah, 2014, 'The dark side of globalization: why Seattle's 1999 protesters were right'. *The Atlantic,* Jan 6, 2014. https://www.theatlantic.com/busi-

ness/archive/2014/01/the-dark-side-of-globalization-why-seattles-1999-protesters-were-right/282831/
Smith, Zadie, 2016, 'Fences: a Brexit diary". *The New York Review of Books*, 13.
Sockness, Brent W., 2004, 'Schleiermacher and the ethics of authenticity: the *Monologen* of 1800'. *Journal of Religious Ethics,* vol.32(3), pp. 477-517.
Sommerville, C. John, 1992, *The Secularization of Early Modern England: From Religious Culture to Religious Faith*. New York : Oxford University Press.
Springsteen, Bruce, 2016, *Born to Run*. New York : Simon and Schuster.
Stasiulis, Daiva & Darryl Ross, 2006, 'Security, flexible sovereignty and the perils of multiple citizenship'. *Citizenship Studies* 10(3), pp. 329-348.
Stein Robert M., 1998, 'Making history English. Cultural identity and historical explanation in William of Malmesbury and Lazamon's *Brut*'. In: Sylvia Tomash and Sealy Gilles eds., 1998), pp. 97-115.
Steinberg, Justin, 2014, *Dante and the Limits of the Law*. Chicago : University of Chicago Press.
Storey, David, 2001, *Territory: The Claiming of Space*. Harlow : Pearson Education Limited.
Storrs, Christopher (Ed.), 2009, *The Fiscal-Military State in Eighteenth-Century Europe: Essays in Honour of P.G.M. Dickson*. Farnham: Ashgate.
Strayer, Joseph, 1971, *Medieval Statecraft and the Perspective of History. Essays by Joseph R. Strayer*. Princeton : Princeton University Press.
Sullam, Simon Levis, 2008, 'The Moses of Italian Unity: Mazzini and nationalism as political religion'. In: Christopher A. Bayly and Eugenio F. Biagini, *Giuseppe Mazzini* etc., pp. 107-124.
Suslov, M.D., 2015, "Holy Rus": the geopolitical imagination in the contemporary Russian Orthodox Church. *Russian Social Science Review* 56(3), pp. 43-62.
Swales, Martin, 1986, 'The problem of nineteenth-century German realism'. In: Nicholas Boyle and Martin Swales, *Realism in European Literature*, pp. 68-84.
Swyngedouw, Erik, 1992, 'The Mammon Quest. "Glocalization", Interspatial Competition and the Monetary Order: The Construction of New Scales'. In: Michael Frederick Dunford and Grigoris Kafkalas (eds.), *Cities and Regions in the New Europe*, pp. 39-67.
Swyngedouw, Erik, 2004, 'Globalisation or "glocalisation"? Networks, territories and rescaling'. *Cambridge Review of International Affairs* 17(1), pp. 25-48.
Taylor, Blair, 2013, 'From alterglobalization to Occupy Wall Street. Neoanarchism and the new spirit of the left'. *City* 17(6), pp. 729-747.
Taylor, Peter and Ben Derudder, 2016, *World City Network: A Global Urban Analysis*. London : Routledge.
Taylor, Peter J., 2002, 'Metageographical moments: a geohistorical interpretation of embedded statism and globalization', in Mary Ann Tetreault et al., *Rethinking Global Political Economy*.

Taylor, Peter J., 2001, 'Visualising a new metageography: explorations in world-city space'. In: Gertjan Dijkink & Hans Knippenberg (eds.), *The Territorial Factor*, pp. 113-128.
Taylor, Peter J., 1995, 'Beyond containers: internationality, interstateness and interterritoriality', *Progress in Human Geography* 19(1), pp. 1-15.
Taylor, Peter, 1995, *The way the modern world works. World hegemony to world impasse.* Chichester : John Wiley & Sons.
Tetreault, Mary Ann, Robert E. Denemark, Kenneth P. Thomas and Kurt Burch, 2002, *Rethinking Global Political Economy. Emerging Issues, Unfolding Odyssees.* London : Routledge.
Tilly, Charles, 1990, *Coercion and Capital in European States, 900-1990*. Cambridge (Mass.) : Blackwell.
Toffler, Alvin, 1970, *Future Shock*. New York : Random House.
Tomasch, Sylvia and Sealy Gilles, 1998, *Text and Territory*. Philadelphia: University of Pennnsylvania Press.
Tracy, James D., 1978, *The politics of Erasmus*. Toronto : University of Toronto Press.
Tunander, Ola, 2004, 'Securitization, Dual State and US-European geopolitical divide or: the use of terrorism to construct World Order'. Paper presented at the Fifth Pan-European International Relations Conference (Panel 28 Geopolitics), The Hague, 9-11 September 2004.
Turam, Berna, 2015, *Gaining Freedoms. Claiming Space in Istanbul and Berlin.* Stanford : Stanford University Press.
Turnbull, Stephen, 2003, *The Ottoman Empire 1326-1699*. Oxford : Osprey Publishing.
Urbinati, Nadia, 2008, 'The legacy of Kant: Giuseppe Mazzini's Cosmopolitanism of Nations'. In: Christopher A. Bayly and Eugenio F. Biagini, *Giuseppe Mazzini* etc., pp. 11-35.
Vandergeest, Peter and Anusorn Unno, 2012, 'A new Extraterritoriality? Aquaculture certification, sovereignty and Empire'. *Political Geography* 31(6), pp. 358-367.
Voegelin Eric, 1940-1941, 'The Mongol Orders of Submission to European Powers, 1245–1255', *International Journal of Byzantine Studies* (American series 1), XV.
Vogt, Henri, 2005, *Between Utopia and Disillusionment: A Narrative of the Political Transformation of Eastern Europe.* New York: Berghahn.
Wakefield, Andre, 2009, *The Disordered Police State: German Cameralism as Science and Practice.* Chicago: The University of Chicago Press.
Waldron, Arthur, 1990, *The Great Wall of China. From History to Myth.* Cambridge : Cambridge University Press.

Walls, Michael and Steve Kibble, 2013, 'Identity, stability and the state in Somaliland: indigenous forms and external interventions'. In: Emma Leonard and Gilbert Ramsay Eds., *Globalizing Somalia*, pp. 253-277.
Watson, Adam, 1992, *The evolution of international society*. Routledge: London.
Watt, W. Montgomery, 1953, *Muhammad at Mecca*, Oxford : Clarendon Press.
Webb, Diana, 1996, *Patrons and defenders. The saints in the Italian city-states*. London : I.B. Tauris.
Weinbrot, Howard D., 1993, *Britannia's Issue. The Rise of British literature from Dryden to Ossian*. Cambridge : Cambridge University Press.
Weizman, Eyal, 2007, *Hollow Land. Israel's Architecture of Occupation*. London : Verso.
Wells, Charlotte C., 1995, *Law and Citizenship in Early Modern France*. Baltimore : The Johns Hopkins University Press.
Wheatley, Jonathan, 2005, *Georgia from National Awakening to Rose Revolution. Delayed Transition in the Former Soviet Union*. Aldershot: Ashgate.
Whited, Tamara, 2000, *Forest and Peasant Politics in Modern France*. Yale University Press: New Haven.
Williamson, Arthur, 1999, 'Patterns of British identity. Britain and its rivals in the sixteenth and seventeenth centuries'. In: Glenn Burgess ed., *The New British History* etc., pp. 138-173.
Wilson, Andrew, 2005, *Virtual Politics. Faking Democracy in the Post-Soviet World*. New Haven : Yale University Press.
Winock, Michel, 2001, 'Sartre: l'effet de modernité'. In : Ingrid Galster, *La Naissance du Phénomène Sartre*, 2001, pp. 200-212.
Worthington, Ross, 2003, *Governance in Singapore*. London : Routledge-Curzon.
Woude, A.M. van der, 1980, 'De alfabetisering'. In: *Algemene Geschiedenis der Nederlanden* 7, Haarlem : Fibula – van Dishoeck, pp. 257-264.
Wright, Derek ed., 2012, *Emerging Perspectives on Nuruddin Farah*. Asmara : Africa World Press Inc.
Yang, Lien-sheng, 1968, 'Historical notes on the Chinese world order'. In: John K. Fairbank, *The Chinese World Order*, pp. 20-33.
Zamoyski, Adam, 1992/1997, *The Last King of Poland*. London : Weidenfeld & Nicholson.
Zapalac, Kristin E.S., 1990, *In His Image and Likeness': Political Iconography and Religious Change in Regensburg, 1500-1600*. Ithaca : Cornell University Press.
Zwitzer, H.L., 1991, *De militie van den Staat. Het Leger van de Republiek der Verenigde Nederlanden* [The Militia of the State. The Army of the Republic of the United Netherlands]. Amsterdam : van Soeren & Co.

INDEX

1..10 Subject mentioned on each page from 1 to 10
1–10 Pages 1-10 cover coherent discussion on subject.

Abkhazia, 176, 180, 182, 184..186
accountability, 39, 125, 136, 145, 147
active (territorial order), 7
Afghanistan, 46, 136, 187
Africa, 3, 19, 46, 49, 131, 148, 152, 154, 169, 184, 189, 191, 194, 196, 197, 199, 204, 210, 231, 237, 239, 240, 244
African Union, 193
Agnew, John, 20, 21
Ahmed, Akbar, 10, 195
Aideed, Mohammed Farah, 193
Al Qaeda, 10, 126, 152, 195, 198, 199
Alexievich, Svetlana, 185
Algeria, 202
Alsace and Lorraine, 162, 165
alter-globalist, 215
Althusius, Johannes, 69
Amalrik, Andrej, 173
anarchism, 208
  and neo-liberalism, 214, 215
Anderson, Benedict, 118
Apple, 209
Arab world, 11, 197
Arezzo, Guitone d', 56
army, ii, 5, 6, 17, 25, 27, 29..33, 36, 38..40, 42..44, 46..49, 51, 65, 66, 69, 74..76, 78, 92, 95, 105, 123, 127, 128, 130, 134, 141, 147, 153, 161..164, 168, 170, 172, 173, 177, 178, 182, 185, 186, 191..193, 197, 198, 201, 210, 223
  military farming, 33
Artois, 73
ASEAN, 138

Asia, i, 1, 35, 46, 50, 122, 138, 140, 171, 173
  Asian values, 137
Asian Regional Forum, 138
Austria(n), 67, 101, 111, 119, 129, 130, 157
Azerbaijan, 176, 179, 184
Bacquet, Jean, 53, 81, 85
Baldwin, Richard, 21, 209
Balkans, 28, 30, 45, 49, 119, 183, 184
Baltic republics / states, 160, 161, 173, 175, 176, 177, 182
barbarians, 5, 9, 25, 27, 29, 32, 35, 36, 37, 40, 41, 61, 73, 154, 202
Barber, Benjamin, 220, 221, 222
Barre, Siad, 191..193, 200, 202, 203
Batam, 139, 140
Bavaria, 89, 130
Bayart, François, 21, 22, 24
Bayezid I (Sultan), 44
Beck, Ulrich, 123
Beissinger, Mark, 174, 175, 177
Belarus, 183
Bell, David, 103
Benelux, 171, 202
Bidassoa river, 73
Bintan, 139
Blanning, T, 93, 104, 105
Blut und Boden, 95, 186
Bocaccio, 57
Bodin, Jean, 42, 69, 78, 80, 81
Boko Haram, 152
Bologna, 57
Borama, 204
border control, 184, 217

246   INDEX

Bosnia, 30
bounded, 2, 7, 126
Brazil, 3, 108, 109, 131
Brexit, 130, 131, 211, 216
BRICS, 3
Britain, 46, 67, 72, 80, 83, 111, 116, 130, 161, 164, 166, 171, 202
  British Empire, 26, 46, 122, 164, 165
Browder, William, 142
Brown, Peter, 39, 40
Brubaker, Rogers, 178
Bull, Hedley, 5, 18, 19, 27
Bullinger, Heinrich, 69
bundling (of authority), 20, 60, 61, 63, 216
Bush, George W., 14, 152
Byzantine Empire, 10, 11, 41, 45
Caesar, Julius, 12
Caliph, 49
caliphate, 12, 41, 152, 215, 218
cameralism, 82, 83, 105, 110, 111
Cameroon, 149
Carré, John le, 15
Catholics, 3, 63, 67, 71, 72, 104, 109, 132, 169, 172
Caucasus, 49, 173, 177
Celtic, 73
Central European University, 126
Cerdagne, 75, 76, 78
CETA, 211
Chad, 149
Chanel, Coco, 167
Charlemagne, 27, 40
Chechnya, 176, 183
Chernyshevsky, Nikolai, 57
China, 3, 8, 14, 16, 27, 29, 30–37, 40, 47..50, 126, 131..133, 137, 138, 141, 145, 147, 155, 173, 181, 209, 223
Chinggis Khan, 25, 48, 50

Chisinau, 178, 183, 184
Christchurch, 2
Christianity, 10, 12, 13, 23, 28, 38..41, 44, 45, 50, 54, 55, 59, 60, 71, 83, 87, 92, 98, 104, 116, 159, 170, 188, 197, 198, 210, 217
Church, 3, 28, 40, 58, 61, 63, 69, 78, 84, 86, 103, 108, 109, 117, 167, 168, 188, 189, 201, 202, 217
Churchill, Winston, 166
CIA, 126, 135
CIS (Commonwealth of Independent States), 181, 182
cities
  collective control neighbourhoods, 220
  postwar urban transnationalism, 172
  escape from Vichy mindset), 168, 169
  Italian, 56–60
citizenship (incl. principles), 18, 78, 81, 86, 93, 113, 136, 167, 182
clans, 192, 195, 196, 199, 200, 203..205
  cutting through clans, 203
Clark, Christopher, 93
Clément, Caty, 175, 190, 194
Cold War, 127, 191, 192, 203
colonialism, 16, 18, 45–46, 199
  neo-colonialism, 18, 209
Columbian age, 121, 122
Commons, 71
Conflent, 75
consociational politics, 130
Constantine (Emperor), 39
consulate, 56
contado, 60, 80
Convergence, Great, 21, 209, 212
Corbières, 74

corruption, 51, 58, 122, 126, 133, 142, 145, 146, 150..152, 156, 181, 200, 211, 216
cosmopolitanism, 27, 115, 200, 202
covenant, 68, 69
crime, 19, 55, 83, 107, 108, 126, 142, 143, 153, 180, 184
Crimea, 49
Crimean Khanate, 43
Croatia, 30, 43
Cromwell, Oliver, 71
Crone, Patricia, 11
Crossan, John Dominic, 12
cross-border cooperation, 17, 81, 111, 141, 142, 172, 203, 221–22
Cultural Revolution, 132
cyber-war, 14, 201
Cyprus, 210
D'Aguesseau, Henri-François, 103
Dabiq, 123
Dandolo, Andrea (Doge), 59
Dante, 51, 52, 86
Danube, 30
Darood (clan), 195
Darwin, 116
David, King, 54
Davies, Thomas, 127
De Gaulle, Charles, 165, 169, 172, 202
Deleuze, Gilles, 5, 157, 159, 165, 195
democracy, 14, 15, 17, 54, 58, 85, 95, 108, 115, 123, 126, 127, 130n, 131, 134..136, 142, 155, 161, 163, 166, 169, 177, 197, 207, 208, 212, 213, 217
Denemark, Robert A., 112
Deng Xiaoping, 132
despotic power, 107
de-territorialization, 49, 54, 96, 97, 117, 158, 160, 169, 177, 195, 201, 204, 205

diaspora, 6, 16, 132..134, 141, 184, 198, 200, 203, 205
diplomacy, 46, 73–79, 86, 112, 124
  absence of symmetry, 48
  para-diplomacy, 124
Divergence, Great. *See* Convergence
Djibouti, 190, 197
dominion, 47
Duffy, Rosaleen, 149
Dugin, Aleksandr, 179, 186..188, 219
Dutch anthem. *See* Wilhelmus
Dutch nation, 54
Dutch Republic, 53, 54, 64, 66, 67, 70, 80, 81, 84
Dutch state (WW II), 163
Duterte, Rodrigo, 217
earthquake, 1, 2, 3, 164, 208
  geopolitical, 2–4
eco-certification (MSC, FSC), 212–13
ecological threat, 110, 117, 207, 156, 212, 213
education, general, 78, 92, 93, 100..102, 107, 113, 114, 118, 150, 167, 176, 215, 216, 220
Egal, Muhammad, 191, 203
Egypt, 10, 136. 107, 218
Eliot, T.S., 51
Elizabeth I, 71
empire
  evil, 61
  imperial overstretch, 35
  imperial trauma, 11
  pride in Soviet Union, 179
  semantics, 25–26
  vs. state, 26–27
England, 15, 31, 41, 54, 55, 63, 71, 72, 80, 83, 104, 151, 163, 165, 184,
English nation, 54
Engvall, Johan, 144, 181
Enlightenment, 85, 95, 96, 103

Erasmus, Desiderius, 61, 70, 82, 85, 92
Erdoğan, Recep, 217
Estonia, 160, 182, 183
Ethiopia, 190, 191..193, 197
ethnic diversity / tension, 5, 19, 29, 30, 33, 44, 45, 48, 72, 81, 99, 101, 137, 138, 141, 142, 152, 154, 160, 175–180, 182..188, 194, 196, 199..201, 208, 219
EU. *See* European Union
Eurasianism, 186, 188, 219
Euro currency, 210
Europe, 2, 4, 5, 7.. 9, 14, 15, 21, 23, 27..30, 37, 38, 40, 41, 44, 46, 47, 49, 50, 51, 53, 56, 60, 61, 64, 65, 67, 72, 79..81, 91..94, 98..101, 104, 105, 107, 108, 111, 119, 121, 122, 129, 135, 140, 152, 153, 157..159, 160, 163, 164, 167, 169, 170..173, 175, 181, 183, 184, 200, 208, 210..212, 221
  influence of, 46
  Council of, 135
European Union, 15, 19, 124, 128, 135, 140, 159, 171, 202, 209, 210, 218, 221, 222
  antagonism between members, 210–11
  constitution, 210
  Court of Justice, 15, 128, 130
  European Parliament, 128, 135, 136
existentialism, 167
extraordinary rendition, 14, 126, 135
extra-territoriality, 16, 17, 127, 138
Falun Gong, 132
Farah, Nuruddin, 193, 199, 200, 203
Fashoda incident, 46
Ferrante, Elena, 1
feudal system, 37, 38, 42, 57, 80

fief, 38, 44
Finland, 189
Florence, 51, 57
foreigners, 9, 33, 38, 41, 43, 44, 81, 106, 155, 182
Foucault, Michel, 219
Fourth Political Theory, 187, 188
France, ii, 21, 25, 28, 41, 44, 46, 50, 53, 60, 61, 63, 66, 67, 70, 71, 73..82, 86, 91, 94, 95, 100, 103, 104, 111, 113, 116, 117, 119, 129, 130, 145, 157, 159, 161..172, 190, 197, 202, 204, 210, 217
Friedrich, Caspar David, 97, 117
FRONTEX, 128
frontier, 5, 22, 25, 28..33, 40..43,, 46, 50, 66, 67, 75, 80, 128, 171, 210
  American frontier, 29
  and field, 22
fuero, 41
G7, 144, 145, 147, 209
Gamsakhurdia, Zviad, 180, 181
Gates, Bill, 124
Gellner, Ernest, 28, 100, 101, 102, 116
gems, 149, 154
Georgia, 144, 145, 154, 176..178, 180..182, 184, 186, 202,
Germany, ii, 15, 27, 41, 46, 56, 61..63, 73, 79, 80, 82, 83, 89..97, 104..106, 109, 111, 116, 119, 123, 129, 130, 145, 161..166, 168..173, 187, 202, 210, 217
  post-war plans about its future, 170–72
gerrymandering, 176, 179
ghaza/ghazi, 45, 49
Githongo, John, 150, 151
glasnost, 174, 175, 177
global framing, 198
Global South, 208, 209, 212

INDEX 249

See also Third World
global street, 219, 220
globalization, ii, 3..5, 10, 18, 20, 21, 23, 46, 122, 127, 153, 154, 185..188, 198, 200, 201, 204, 207, 209, 210..213, 216..220, 222, 223
glocalization, ii, 16, 21, 24, 153, 155, 203, 205, 213, 214, 219, 220, 222
Goethe, Johann Wolfgang von, 95
Gonzalez, José Grinda, 143
Gorbachev, Mikhail, 174, 175, 177
governance, good, 61, 151, 200, 203, 215
Greece, 210
Greenfeld, Liah, 91, 116
Grenz, Stanley J., 116
Guattari, Felix, 5, 157, 159, 165, 195
Gumilev, Lev, 186
Habsburg Empire, 28, 30, 44, 54, 65, 66, 80
Han (dynasty), 25, 29, 35, 36, 47, 48
Hardt, Michael, 19, 20
Hargeisa, 192, 196
Haro, Don Luis de, 73
Hawiye (clan), 194
hegemony, 3, 46, 47, 65, 106, 137, 138, 202, 213
Heidegger, Martin, 187, 166
Henne, Peter, 198
Herzen, Alexander, 187, 219
Hesse, 79
Hitler, Adolf, 91, 157, 161, 163, 164, 173, 202
Hoffmann, Matthew J., 112
holey space/territory, 159
Holland, 65, 66, 67, 104, 118
Holy Roman Empire, 15n, 41, 60, 79, 80, 129
homicianos, 41
human rights, 125, 128, 130, 174, 196, 217

Hungary, 43, 101, 126, 211, 217
Huntington, Samuel, 10
Hussein, Saddam, 193
identity politics, 154, 183
illiberal trends, 126, 147, 202, 211, 219
Ilyin, Ivan, 219
India, 3, 155, 189, 217
Indonesia, 138..140, 189
Industrial Revolution, 46, 101, 104, 174, 209
industrialisation, 92, 102, 103, 106
information war. See cyber-war
infrastructural control/power, 6, 9, 23, 92, 95, 100, 107..109, 115, 156
INGO, 125..127, 131–33, 148, 149
inter-stateness, 127
inter-territoriality, 127
invented tradition, 84
Iraq, 141, 152, 158, 187
Isaac (clan), 62, 63, 192..195, 197, 203
Islam, 10, 11..13, 23, 24, 27, 28, 41, 44, 45, 49, 149, 152, 193..195, 197, 198, 199, 218
Islamic State, 3, 5, 123, 152–53, 158, 161, 218
Islamic world, 10, 49
  Muslim lands, 198
Israel, 11, 17, 18, 53, 60, 68, 71, 81, 88, 155
Italy, 1, 2, 52, 55, 56, 58.. 60, 64, 79..81, 100, 109, 111, 113, 115, 119, 123, 144, 145, 152, 162, 190, 195, 217
Jackson, Julian, 170
James VI, 72
Jamet, Claude, 164, 202
Japan, 46, 109, 144, 171, 212
Jerusalem, i, 50, 55, 56, 58, 86, 87
Jesus, 12, 207

Jews, 12, 13, 28, 60, 64, 68, 71, 73, 91, 169
jihad, 44, 153,
Jones, Philip, 58
Justi, Johann von, 83
Kabila, Laurent, 150
Kant, Immanuel, 95
Kenya, 150, 151, 190, 195, 197, 198
Khadaffi, Muammar, 193
Kibaki, Mwai, 151
kingdom (term), 12
kings, 7, 9, 38, 40..42, 47, 50, 54, 55, 60, 61, 63, 71, 72, 81
Kirill, Patriarch, 188
korenizatsiya, 176
Kyrgyzstan, 144
land grab, 16–17
language (as territorial marker), 55, 92, 94, 98, 100, 102, 109, 115, 119, 175, 176, 182, 184, 210
Latin, 12, 38, 55, 69, 94, 101
Latvia, 182
Laval, Pierre, 166, 169
law (rule of), 146
Lebanon, 175, 190
Lebrun, Albert, 163
Lee Kuan Yew, 136
legible, 26, 125, 126, 127, 131, 154, 223
legitimacy, ii, 9, 41, 42, 49, 52, 53, 59, 64, 71, 81, 95, 105, 113, 131, 150, 152, 163, 165, 166, 169, 177, 180, 190, 192, 194, 195, 196, 213
    territorial vs. political, 166
Levant, 152
Li Bai, 25, 29, 44
Li Hongzu, 132
line of flight, 157, 158, 159, 160, 165, 172, 201, 203, 205
List, Friedrich, 89, 106, 115
Loriaux, Michel, 171

Louis II, 60
Louis VI, 60
Louis XIV, 73, 103
Loveman, Mara, 108
Luther, Martin, 61
Mackinder, Halford, 121, 122, 179
Macpherson, James, 98, 102
Madagascar, 149
mafia, 143, 144
Magnitsky, Sergei, 142, 143
Magritte, René, 26
Maherteen (clan), 202, 203
Maier, Charles, ii, 22, 82
Malaysia, 136, 138, 140
Malmesbury, William of, 54, 55
Manon Lescaut, 94
Mao Zedong, 132
Marco Polo, 31
Maria-Theresa, 73
Martines, Lauro, 55, 56, 57
Mastersingers of Nuremberg, 89–91
Mazarin, Cardinal, 73
Mazzini, Giuseppe, 100, 113..115, 116, 118
medievalism. *See* neo-medievalism
Melucci, Alberto, 123
Menkhaus, Ken, 189, 194
meritocracy, 137
Mexico, 211
Microsoft, 124
migrants, 30, 101, 131, 140, 209, 217, 218
military. *See* army
millennialism, 188, 219
Ming (dynasty), 29, 33, 50
Minzner, Carl, 133
MIT Industrial Performance Center, 109
modernization, 91, 99..102, 176
Modi, Narendra, 217

Mogadishu, 189, 193, 198..200, 203, 205
Mohamoud, Abdullah, 192
Moïsi, Dominique, 18
Moldavia, 43, 178, 179
Molotov-Ribbentrop pact, 161
Monnet, Jean, 171
Möngke Khan, 25, 48
moral indoctrination, 58
moral principle, i, ii, 3, 6, 13..15, 22, 24, 27, 39, 46, 58, 61, 64, 79, 84..86, 88, 91..94, 111..114, 116, 118, 155, 156, 167, 172, 185, 189, 195, 200..202, 204, 205, 208, 209, 215..220, 222, 223, 225, 226
   Asian values, 137–38
Morgenthau, Hans J., 134
mosque, 149
Müller, Adam, 106
multinational firms, 124, 213
multilateral cooperation/treaties, 111, 112, 113, 113, 115, 119, 141, 213
music (spirit of a time), 90, 91, 151, 160, 161, 201, 202, 223
NAFTA, 211
Nagorno-Karabakh, 179, 184
Napoleon, Bonaparte, 95, 96, 119, 130, 173
national identity, 26, 55, 70, 90, 99, 100, 105..107, 109, 110, 115, 130, 178, 185, 195, 199, 210, 213, 216
   and European Union, 210
nationalism, 6, 9, 45, 55, 68, 91, 92, 95, 97, 113, 115, 116, 118, 156, 160, 171, 172, 180, 186, 211, 215..217, 222, 223
   duty, 114
   economic constraints, 109–10
   economic nationalists, 105
   genetic (germany), 95
   identity, 100
   imagined community, 118
   industrialisation (Germany), 91
   romanticism, 96
   theory, of 99–107
nationality politics Soviet Union, 175, 178–79
Needham, Joseph, 31, 36, 37
Negri, Antonio, 19, 20
neo-colonialism (fear of), 211–13
neo-medieval, 19, 20
neo-nationalism, 156, 216
Netherlands, i, 66, 67, 70, 104, 130n, 134, 163, 210
Newton, 116
NGO, 131–33, 213
Nigeria, 148, 151
nobility, 45, 46, 57, 79, 82, 103, 116, 157, 226
Norman Conquest, 55
Norway, 189
novel (prefigures new age), 15, 57, 93, 193, 199, 200
oblast, 176, 180
Occupy Wall Street, 207, 208, 214, 220
Ogaden, 190..193, 199
okrug, 176
oligarchy, 147
Olson, Roger, 116
Orban, Viktor, 217
Orthodox Church, 28, 188, 189, 201, 202
OSCE, 182
Ossian, 72, 98, 99, 102, 171
Otto I, 28, 41
Ottoman Empire, 27..29, 43–45, 49, 50, 101, 111, 119
outsourcing production, 4–5, 208, 209
Pan Ku, 36
Panama Papers, 124

PAP (People's Action Party), 137
Patriotism, 56, 67, 103, 169
Paul (apostle), 12
Pentecostal Church, 22, 217
perestroika, 174
Persian Empire. *See* Sassanid Empire
Pétain, Henri-Philippe, 163..169
Philip IV (Spain), 73
Phillips, Sarah, 203
philosemitism, 71
Piedmont-Sardinia, 119
pietism, 92
plague, 75
Poland, 41, 83, 119, 157, 158, 211
police, 20, 82, 107, 108, 128, 135, 143, 144, 167, 177, 178, 194, 200, 215, 220
policing, 15, 107, 151
pope, 3, 9, 37, 40, 48, 49, 50, 52, 55, 56, 60, 61, 87, 132, 225
popolo, 56, 57
populism, 217, 230
Portugal, 210
Prévost, Abbé, 94
private (non-state), 3, 58, 83, 105, 123, 124, 125, 126, 127, 135, 148, 152, 153, 159, 170, 177, 205, 211
  illegible, 148–52
Protestants, 63, 65, 71, 72, 79, 92
proxy wars, 123
Prussia, 41,78, 80, 92..94, 104, 106,, 109, 110, 116, 119, 130, 157
public interest
  clientelism, 191
  public opinion, 111, 124, 147, 171, 181, 186, 201
  public order, 42
  public space, i, 56, 89, 97, 105, 117, 208
  public sphere, 93, 95, 104, 105, 135

Puigcerda, 75, 221
Puntland, 195
Putin,Vladimir, 5, 144, 185
Pyrenees, 40, 73, 74, 78, 81, 162, 196, 221
  peace of, 73–78
Qi gong, 132
Qin (dynasty), 31, 33, 48
Quraysh (tribe), 49
Rajaratnam, Sinnathambi, 138
Reagan, Ronald, 174
rebels, 48, 117, 125, 148, 151, 185, 186, 192
Red Cross, 126
Reed, Jonathan L., 12
Reformation, 9, 63, 70, 86, 92, 103, 215
refugees, 2, 17, 30, 70, 192, 193, 210, 211
regeneration. *See* self-transformation
Reichskreise, 129
Reichstag, 129
religion, 215
  European supranationalism, 172
  Exodus tale, 60, 64–73
  state denial, 158
  fusion with politics, 58–61
res publica, 60, 61
rescaling, 213, 219, 242
ressentiment, 116
re-territorialization, 5, 23, 24, 53, 96, 139, 157, 158..160, 179, 195, 201, 204, 205, 216
Reuter, Paul, 170
Reynaud, Paul, 163
Ring (opera), 91
Ringmar, Erik, 79
Roman Empire, 8, 12, 13, 15n, 24, 27, 31n, 37, 39, 40, 44, 48, 53, 61, 81, 170, 172, 215
Romania, 179, 183, 211

Ross, Darryl, 136
Rousseau, Jean-Jacques, 90
Roussillon, 73, 74, 78
Rovan, Joseph, 157
Ruggie, John, 20, 21, 42
rule of law, 177
Rushdie, Salman, 141
Russia, 3, 28, 50, 57, 111, 126, 141..145, 154, 157, 159, 173, 175..178, 180..189, 197, 201, 204, 205, 219
saints, patron, 58..60
sanjak, 44
Sartre, Jean-Paul, 167..169, 172
Sassanid Empire, 10, 11
Sassen, Saskia, 21, 220
Sava, 30
Savar (Bangladesh), 208
Savoy, 117, 119
sblizheniye, 176
Schama, Simon, 67, 68, 70
Schleiermacher, Friedrich, 97, 116, 117
Schuman, Robert, 171
Scotland, 72, 80, 98
Scott, James, 26, 110, 111
Seattle demonstrations, 207, 208, 212, 214
secret prisons, 14, 125, 126
sejm, 157
self-awareness, 57
self-determination, 54, 79, 175
self-government, 54, 79, 203
self-realization, 116
self-transformation, 40
   national rebirth, 164
   purification, 168
   regeneration, 97, 104, 159
   revitalization of Europe, 172
September-11, 4, 5, 10, 134, 208
Sepúlveda, 41

Serbia, 30
Shaba(a)b, Al-, 152, 193, 198
shadow economy. *See* underground economy
sharia, 198, 218
Shell, 148
Shevardnadze, Eduard, 181
Showalter, Daniel, 12
Silk Road, 181
Singapore, 136, 137, 138..141, 154
singing revolution, 160–61
slaves, 38, 44, 101, 108, 109
sliyaniye, 176, 178
Slovakia, 211
SMI (growth triangle), 139–40
Smith, Adam, 85
SNM (Somali National Movement), 193
Snowden, Edward, 15
Somalia, 159, 175, 189, 191, 193, 194, 195, 196, 197, 198, 199, 200, 202, 204, 226, 231, 235, 237, 238, 239, 243
   Federal Government, 193
   Trans. Federal Government, 196, 198
   Trans. National Government, 195
Somaliland, 151, 193..197, 200, 203..205, 227
   international partners of, 197
Soranzo, Doge, 59
South America, 45
South Ossetia, 180, 182, 184, 186
sovereignty, 6, 7, 9, 17..19, 37, 42, 63, 69, 76, 78..80, 86, 95, 105, 125, 128..130, 133, 136, 140, 141, 148, 150, 153, 159, 166, 169, 170, 172, 196, 210..212, 221
   flexible, 133–36

Soviet Union, 3, 159, 160, 161, 169, 171..182, 184, 185, 191, 201, 204, 205
  pride in Soviet Union, 185
  secessionist frame, 174
  secessions, 174–85
  transnational tidal force, 174
Spain (Spanish), 41, 54, 65..69, 73..76, 78, 143, 210, 217
Spanish Empire, 68
specialisation, 101, 169
Springsteen, Bruce, ii, 207
Stadholder, 65, 67
Stasiulis, Daiva, 136
state
  centralisation, 23, 39, 50, 51, 64, 65, 72, 78, 81, 94, 194
  collapse, 152, 158, 173, 189, 190, 191, 193, 201, 204
  de facto, 184, 193, 195
  death, 158, 159, 201
  forestry, 110, 117
  legibility, 110, 111
  national unification, 100, 119
  secularization, 40, 70, 84, 87–88, 156, 218
  security, 126, 134
  shadow state, 126, 148, 149, 154
  state-building failure, 194
  symbolic power, 93, 108–9
States General, 64
subjectivation, 22, 23
sultan, 43, 44, 49, 50, 101
supranational governance, 128–31, 169
supranationalism, 15, 124, 125, 155, 171, 205, 209 223
survivalism, ideology of, 137
suzerainty, 47
Sweden, 80, 86n, 129, 197
Swift money transfer, 135

symbolic multiplier, 123, 125, 205
Syria, 136, 158, 187
szlachta, 157
Taliban, 152
Tambogrande, 213
tariffs, 106, 209, 213
Tatar(stan), 28, 175, 176, 179
taxation, 6, 26, 82, 86, 92, 105, 128
  fiscal-military state, 105, 118
Tbilisi, 177, 180, 182, 186, 237
territorial order, i, 5, 7, 8, 13, 15, 18, 21, 23, 24, 50, 51, 57, 80, 154, 155, 159, 171, 205, 215, 223
terrorism, 14, 18, 126, 135, 154, 184, 189, 195, 215, 221,
  war on terror(ism), 14, 126, 155
terrorist attacks, i, 4, 10, 153
terrorists, 10, 14, 122, 143, 159
Teutonic Order, 41
Thailand, 212
Third Republic (France), 166–167, 168
Third World, 128, 191, 192 *See also* Global South
torture, 126, 135
Trans-Dniester, 183, 184, 237
transnational (networks), 8, 12, 15, 20, 27, 99, 109, 115, 117, 122, 125, 126, 131..133, 141, 148, 152, 154, 171, 172, 174, 180, 197..199, 200..202, 204, 205, 209, 210, 212, 218
transnational regime, 125, 126, 154
transnational region, 136–41
  fragmented integration, 141
Transsylvania, 43
tribal / tribe, 5, 6, 10..12, 19, 31, 33, 48, 49, 68, 81, 148, 190, 195, 199, 200, 203, 215, 218
tribute, 33, 34, 35, 43

INDEX 255

Trump, Donald, 5, 124, 152, 207..209, 217
TTIP, 211
Tubman, Winston, 195, 196
Tunku Abdul Rahman, 136
134, Turan, Berna, 220
turbulence, 8, 9, 131, 205, 216
Turkey, 210, 217, 218
Tuscany, 56
Uganda, 198
Ukraine, 145, 154, 173, 176..178, 182..186, 202, 211
Umayyad Empire, 12, 28
UN. See United Nations
unbounded. See bounded
unbundling. See bundling
underground economy, 125, 145, 147, 148, 151, 154, 191
United Arab Emirates, 197
United Kingdom. See Britain, England
United Nations, 20, 121, 128, 196
  Committee on Human Rights, 182
  resistance against secessionism, 195
United States (US), i, 2..4, 15, 20, 21, 47, 110, 111, 134, 135, 141, 143, 145, 147, 150, 152, 154, 155, 165, 166, 170, 171, 172, 174, 181, 187..189, 191, 193, 197, 198, 208, 209, 211
  American Dream, 2, 207
  American Revolution, 69
Uriage, 170
vassal(s), 9, 35, 38, 39, 43, 154
Venice, 58, 59, 138
vice, 58, 83, 167
Vichy regime/state, 161..170, 172, 202, 204
Vienna Congress, 96, 119
village, non-territorial principle, 76

virtual (fake) politics, 20, 96, 123, 124, 142, 187
Vojna Krajina, 30
Volk, 92
Wagner, Richard, 89, 90, 91, 97, 98, 117
Waldron, Arthur, 29, 31, 32, 33
Wall, Great, 30–33
Wallachia, 43
walls, 2, 8, 31, 32, 56, 80, 155, 156, 213
warlords, 19, 40, 148, 189, 193, 195, 198, 204
Wars of Liberation (Germany), 91, 95, 96
Watson, Adam, 47
Weinbrot, Howard D., 72
West bank (settlements), 17
West, the, 3, 4, 10, 11, 18, 122, 134, 144, 145, 152, 153, 161, 169, 171, 173, 174, 179, 181, 185, 186, 187, 201, 212, 214, 218
Westphalian Peace, 22, 47, 73, 112, 113, 129
Westphalian state, 22, 124, 148, 153
Wikileaks, 124, 142, 143
Wilhelmus, 53, 68
William of Orange, 54, 66
Wilson, Andrew, 123, 124, 142
women, 25, 40, 104, 108, 204
World Wars, 119
  Great War, 116, 127, 169, 175
  Second WW, 46, 159, 163, 182, 202
Worthington, Ross, 137
WTO, 125, 207, 212
Xiong-nu, 32, 35, 36
Yeats, William Butler, 121, 127
Yeltsin, Boris, 180
Yoruba, 151
Yugoslavia, 30, 175, 190
Zaire, 150

Zhirinovsky, Vladimir, 142

Geographie: Forschung und Wissenschaft

Cay Lienau; Hermann Mattes
Griechenlands Nordosten
Eine geographisch-ökologische Landeskunde
Bd. 8, 2018, 358 S., 29,90 €, br., ISBN 978-3-643-14083-8

Hans Gebhardt (Ed.)
Urban Governance, Spatial Planning and Economic Development in the 21th Century China
vol. 6, 2018, 224 pp., 44,90 €, br., ISBN 978-3-643-90418-8

Forum Politische Geographie

hrsg. von Prof. Dr. Paul Reuber (Universität Münster, Federführung), Prof. Dr. Georg Glasze (Universität Erlangen-Nürnberg), Prof. Dr. Olivier Graefe (Université de Fribourg), Prof. Dr. Benedikt Korf (Universität Zürich), Prof. Dr. Julia Lossau (Universität Bremen), Prof. Dr. Annika Mattissek (Universität Freiburg), Prof. Dr. Martin Müller (Universität Lausanne), Prof. Dr. Anke Strüver (Universität Hamburg)
Schriftleitung: Dipl.-Ing. Claudia Schroer

Ursula Meyer
Foncier Périurbain, Citoyenneté et Formation de l'Etat au Niger
Une analyse ethnographique de Niamey
Bd. 16, 2018, ca. 312 S., ca. 39,90 €, br., ISBN 978-3-643-80287-3

Elisabeth Militz
Affective Nationalism
Bodies, Materials and Encounters with the Nation in Azerbaijan
vol. 15, 2018, ca. 200 pp., ca. 24,90 €, pb., ISBN 978-3-643-80278-1

Claudia Gebauer
Changing Climates: Translating Adaptation in|to Rwanda
Bd. 14, 2018, 256 S., 34,90 €, pb., ISBN 978-3-643-90826-1

Hendrik Meurs
Das Herrschaftssystem von Turkmenistan – Mechanismen zum Erhalt der Macht und Inszenierungen zu ihrer Legitimation
Bd. 13, 2016, 806 S., 54,90 €, br., ISBN 978-3-643-12736-5

Mathias Polak
Zwischen Haushalt und Staat
Lokale Water Governance im zentralen Norden Namibias
Bd. 12, 2014, 256 S., 39,90 €, br., ISBN 978-3-643-12623-8

Sebastian Zug
The Gift of Water
Bourdieusian capital exchange and moral entitlements in a neighbourhood of Khartoum
vol. 11, 2014, 336 pp., 29,90 €, pb., ISBN 978-3-643-90452-2

LIT Verlag Berlin – Münster – Wien – Zürich – London
Auslieferung Deutschland / Österreich / Schweiz: siehe Impressumsseite